Java EE 开发技术及应用

主 编 蔡群英 黄镇建

内 容 简 介

本书系统介绍 Java EE 开发的相关技术,包括 Java EE 概述、前端开发技术、JSP 技术、Servlet 技术、JDBC 数据库连接技术、JavaBean 组件技术、Java EE 软件架构模式、Java EE 综合应用开发,以及 Spring 框架、SpringMVC 框架、Mybatis 框架及 SSM 框架整合。本书强调实用性,知识点讲解透彻并配有相应的案例,各阶段也配有综合案例。本书沿着软件架构模式这条主线对各种 Java EE 技术展开介绍,使读者循序渐进地掌握 Java EE 技术并能够进行应用开发。

本书可作为 Java Web 开发技术课程或 Java EE 开发技术课程的教材,也可作为社会培训教材或自学教材。

图书在版编目(CIP)数据

Java EE 开发技术及应用/蔡群英,黄镇建主编.—哈尔滨:哈尔滨工程大学出版社,2021.8
 ISBN 978-7-5661-3168-3

Ⅰ.①J… Ⅱ.①蔡… ②黄… Ⅲ.①JAVA 语言-程序设计 Ⅳ.TP312.8

中国版本图书馆 CIP 数据核字(2021)第 158360 号

Java EE 开发技术及应用
Java EE KAIFA JISHU JI YINGYONG

选题策划	刘凯元
责任编辑	刘凯元
封面设计	李海波

出版发行	哈尔滨工程大学出版社
社　　址	哈尔滨市南岗区南通大街 145 号
邮政编码	150001
发行电话	0451-82519328
传　　真	0451-82519699
经　　销	新华书店
印　　刷	哈尔滨市石桥印务有限公司
开　　本	787 mm×1 092 mm　1/16
印　　张	14.75
字　　数	385 千字
版　　次	2021 年 8 月第 1 版
印　　次	2021 年 8 月第 1 次印刷
定　　价	55.00 元

http://www.hrbeupress.com
E-mail:heupress@hrbeu.edu.cn

前　言

　　Java EE 开发技术是当前主流的 Web 开发技术，技术体系庞大，与前端开发技术的联系紧密，其内容涉及互联网计算机应用相关的架构、方法和技术。目前，已出版的 Java EE 教材较多，主要分为 Java Web 开发技术的教程和 SSH 框架或 SSM 框架（即 Spring + SpringMVC + Mybatis 三个框架）的教程。SSM 框架是当前主流的 Java EE 第三方轻量级框架，是企业级应用软件开发首选的框架，但这类教程要求读者具备一定的 Java Web 基础。

　　针对这种情况，作者通过精心梳理教学内容，整合教学模块，整理教学案例，从 Java Web 基础到 SSM 框架进行介绍，重视软件架构模式，循序渐进地引导学生进行学习。每一个模块都安排了综合案例，使学生学完之后有一个综合实践的机会，也是对模块内容的总结。本书内容设计合理，便于教师根据学生的实际情况和教学课时进行选取，适合作为 Java Web 开发技术课程或 Java EE 开发技术课程的教材。

　　本书主要章节内容如下。

　　第 1 章 Java EE 概述，介绍 Java EE 的发展历程、技术框架、主要技术及开发环境。

　　第 2 章前端开发技术，介绍 HTML 5 + DIV + CSS、JavaScript 技术、jQuery 框架和 Bootstrap 框架，使读者能够快速地构建前端页面和掌握一些常见的与服务器端交互的方法。

　　第 3 章 JSP 技术，详细介绍 JSP 基本语法、内置对象，并通过综合案例，加深对这些技术的理解和运用。

　　第 4 章 Servlet 技术，介绍 Servlet 技术体系结构、Servlet 编程的流程和文件上传。

　　第 5 章 JDBC 数据库连接技术，介绍 JDBC 体系结构、MySQL 数据库操作并通过数据库操作案例完整地展示了 Model1 架构模式。

　　第 6 章 JavaBean 组件技术，介绍 JavaBean 的概念、生命周期及应用。

　　第 7 章 Java EE 软件架构模式，介绍 Model1 架构模式、MVC 架构模式、多层架构模式，以及 EL 表达式和 JSTL 标准标签库。

　　第 8 章 Java EE 综合应用开发，介绍软件开发的流程、规范，并以留言板为例，介绍 MVC 架构模式的实现。

　　第 9 章 Spring 框架和 SpringMVC 框架，初步认识 Spring 框架，介绍 SpringMVC 框架的工作流程、参数传递、数据校验和文件上传。

　　第 10 章 Mybatis 框架，介绍 Mybatis 框架的工作流程、配置文件，以及 SqlSessionFactory 类、SqlSession 类、映射文件的创建及关联映射、动态 SQL。

　　第 11 章 SSM 框架整合，介绍多层架构模式实现整合的步骤及各层的规划和参数的传递，并以图书管理系统为例，介绍部分功能的实现。

  本书具体分工如下:第 2 章和第 5 章由黄镇建编写,其余章节由蔡群英编写。全书由蔡群英拟定提纲并统稿。

  本书的编写是对作者多年教学内容和案例的整理,书中留言板部分的代码是由黄剑颖和陈慧琳两位同学所编写的,在此对她们表示感谢。

  由于作者水平有限,错误和疏漏在所难免,欢迎读者批评指正。

编 者

2021 年 6 月

目 录

第 1 章　Java EE 概述 ·· 1
　1.1　Java EE 的发展历程 ·· 1
　1.2　Java EE 的技术框架 ·· 2
　1.3　Java EE 的主要技术 ·· 4
　1.4　Java EE 的开发环境 ·· 7
　习题 ··· 18

第 2 章　前端开发技术 ·· 19
　2.1　HTML 5 + DIV + CSS ··· 19
　2.2　JavaScript 技术 ·· 32
　2.3　jQuery 框架 ··· 37
　2.4　Bootstrap 框架 ··· 40
　习题 ··· 45

第 3 章　JSP 技术 ·· 46
　3.1　JSP 概述 ·· 46
　3.2　JSP 基本语法 ··· 47
　3.3　JSP 内置对象 ··· 51
　3.4　JSP 综合案例 ··· 65
　习题 ··· 70

第 4 章　Servlet 技术 ·· 72
　4.1　Servlet 概述 ··· 72
　4.2　Servlet 编程 ··· 73
　4.3　文件上传 ··· 79
　习题 ··· 81

第 5 章　JDBC 数据库连接技术 ·· 82
　5.1　JDBC 概述 ·· 82
　5.2　MySQL 数据库操作 ·· 86
　5.3　数据库操作案例 ··· 89

第 6 章　JavaBean 组件技术 ·· 97
　6.1　JavaBean 概述 ··· 97
　6.2　JavaBean 的生命周期 ·· 99
　6.3　JavaBean 的应用 ·· 100
　习题 ··· 103

第 7 章 Java EE 软件架构模式 · 104
- 7.1 Model1 架构模式 · 104
- 7.2 MVC 架构模式 · 107
- 7.3 多层架构模式 · 110
- 7.4 EL 表达式和 JSTL 标准标签库 · 110

第 8 章 Java EE 综合应用开发 · 115
- 8.1 软件开发流程和规范 · 115
- 8.2 留言板系统设计 · 117
- 8.3 MVC 架构模式的实现 · 118

第 9 章 Spring 框架和 SpringMVC 框架 · 147
- 9.1 Spring 框架初识 · 147
- 9.2 SpringMVC 框架概述 · 152
- 9.3 SpringMVC 框架详细介绍 · 157
- 9.4 SpringMVC 数据校验 · 166
- 9.5 SpringMVC 文件上传 · 169
- 习题 · 172

第 10 章 Mybatis 框架 · 173
- 10.1 Mybatis 框架概述 · 173
- 10.2 Mybatis 框架详细介绍 · 178
- 10.3 Mybatis 框架 SQL 映射文件 · 181
- 习题 · 207

第 11 章 SSM 框架整合 · 208
- 11.1 整合的步骤 · 208
- 11.2 图书管理系统功能描述 · 216
- 11.3 各层的规划与参数传递 · 217
- 11.4 部分功能的实现 · 222

参考文献 · 228

第 1 章　Java EE 概述

【本章内容】

- 1.1　Java EE 的发展历程
- 1.2　Java EE 的技术框架
- 1.3　Java EE 的主要技术
- 1.4　Java EE 的开发环境

1.1　Java EE 的发展历程

Java EE(Java Enterprise Edition)是 Java 平台的企业版,其核心是一组技术规范和指南,其提供了一组功能强大的 API,用于开发和部署可移植、健壮、可伸缩且安全的企业级应用程序。目前,Java EE 已成为企业级开发的工业标准和首选平台。

首先,了解一下 Java EE 的发展历程。Java EE 早期的名称是 J2EE,它是 Java 平台的企业版,Java 是 Sun 公司在 1996 年推出的一种新的纯面向对象的编程语言。1998 年 12 月,Sun 公司发布 JDK 1.2 版本和 JSP、Servlet、EJB 规范,并开始使用 Java 2 这个名称。1999 年,Sun 公司将 Java 2 技术一分为三,即标准版 J2SE、企业版 J2EE 和微型版 J2ME,这是 J2EE 的第一个版本,包含了 Web 开发常用的 Web 层、业务逻辑层、表示层和消息服务。2001 年,Sun 公司发布 J2EE 1.3 版本、2003 年发布 J2EE 1.4 版本,此时 J2EE 所提供的组件和功能已非常强大,但由于使用它开发应用程序时需要进行大量的配置和过度理论化的计算模式,开发效率和运行效率都不高,因此使用者并不多。Sun 公司为了改变这种局面,对 J2EE 做了很大的改变,于 2006 年 5 月发布 J2EE 1.5 版本,并改名为 Java EE 5。Java EE 5 提供功能强大的 API 的同时,通过改善系统架构、简化开发和部署的方式,缩短了开发时间,降低了开发难度,提升了软件的性能。简化开发和部署的方式主要是通过使用更多的默认设置和采用标注的形式,使其更方便开发。

其次,Java EE 是一套技术规范和指南,与之相适应的各种开发平台和服务器软件都要遵循相关约定。Java EE 的发展历程如图 1.1 所示,列出了早期的各个版本所包括的主要技术规范。

2009 年 12 月,Sun 公司正式发布了 Java EE 6,包括了 JSF 2.0、JSP 2.1、JSTL 1.2、Servlet 3.0、EJB 3.1、JPA 2.0 和 JAX-RS 1.1 等规范。自 2013 年 6 月 Sun 公司发布 Java EE 7 以来,Java 开发团队一直在规划下一个版本,2017 年,备受期待的 Java EE 8 正式发布,规范

包括了两个新的 API(JSON-Binding 1.0 和 Java EE Security 1.0),并改进了已有的 API(JAX-RS 2.1、Bean Validation 2.0、JSF 2.3、CDI 2.0、JSON-P 1.1、JPA 2.2 和 Servlet 4.0),包含数百个新特性、更新功能和错误修复。

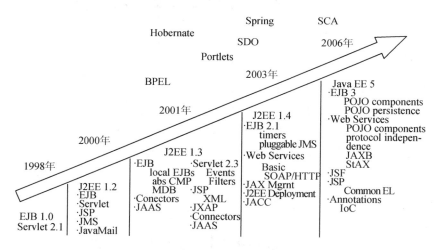

图 1.1　Java EE 的发展历程

1.2　Java EE 的技术框架

1.2.1　企业级应用程序的体系结构

Java EE 作为一个分布式企业应用开发标准,通过它的整个技术框架来规范企业级应用软件的开发。首先,需要了解什么是企业级应用软件。企业级应用软件指的是给大型的企业或组织创建的应用软件,这些企业或组织通常有多个分支机构,分布在不同的位置,各个部门或组织可能有自己的信息系统软件,这些软件所使用的开发语言和运行平台可能不同。对于这样的情形,采用什么技术能够在现有软件系统的基础上实现整个企业的应用软件的开发呢? Java EE 作为分布式企业应用开发标准,以 Java 语言为基础,采用虚拟机机制,使得它具有跨平台的特性;并且 Java EE 的技术体系也使其开发的应用软件能够适应企业的快速反应和可扩展性,能够与企业原来的信息系统集成,具有分布式和极高的安全性。

企业级应用软件,通常采用 4 层体系结构,即客户层、表示层、业务逻辑层和数据库层,如图 1.2 所示。

1.2.2　Java EE 的技术框架

对于企业级应用软件的 4 层体系结构,Java EE 有与之相适应的技术框架。图 1.3 是 Java EE 8 的技术框架。

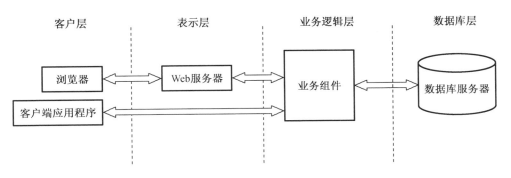

图 1.2　企业级应用软件的 4 层体系结构

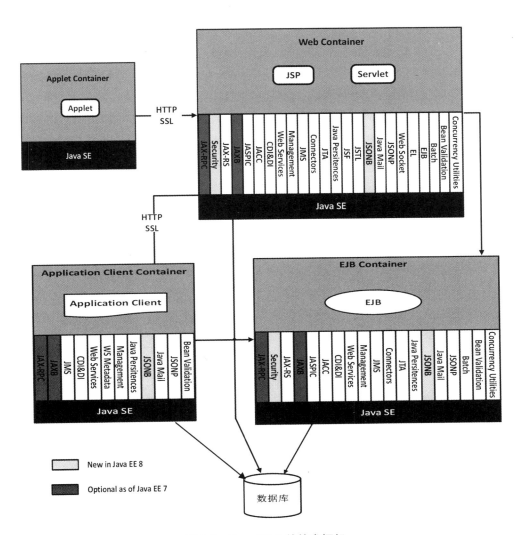

图 1.3　Java EE 8 的技术框架

图 1.3 包括 5 部分,除了数据库外,其余每一个部分里面灰色背景的大矩形块表示的是容器,在矩形块里面的是运行在容器中的组件,最底下黑色背景的矩形块是 Java SE,它是所有 Java EE 技术的基础,在 Java SE 之上的是各种通信技术和服务技术。

在 Java EE 技术框架中，容器是提供给 Java EE 组件运行的环境，容器的开发要遵循 Java EE 规范来实现，其基本接口和功能要遵循规范，具体的实现则由容器厂商来决定，这样才可以做到统一、规范。

对企业级应用软件所划分的 4 层体系结构，Java EE 技术框架提供了 4 层的技术支持。在客户层，提供了实现 B/S 模式和 C/S 模式的技术。对于 B/S 模式，提供给用户的是网页的形式，除了 HTML 网页，嵌入网页的 Applet 程序，就是 Java EE 技术所提供的客户端组件。Applet 程序是 Java 的小程序，它的运行环境是浏览器，是在浏览器里面所安装的 Java 虚拟机上运行的。对于 C/S 模式，客户端应用程序（Application Client）是用 Java 语言编写的，借助 Swing 等图形界面工具创建的桌面应用程序，可以直接访问运行在 Java EE 服务器中的 EJB 组件，实现业务逻辑，它的运行环境是客户端计算机操作系统的 Java 虚拟机。桌面应用程序（也可以是由其他编程语言编写的桌面应用程序），跟 Java EE 服务器进行交互，使得遗留系统能够集成在 Java EE 应用系统中。

对于表示层，组件是 JSP、Servlet。这些组件的编写要依照 Java EE 的技术规范，其运行环境是 Web 容器，常用的 Web 容器是 Tomcat 服务器。Java EE 官方推荐的是 GlassFish 服务器。Java EE 官方提供 SDK（SoftWare Development Kit），并将 GlassFish 一起捆绑下载，始终支持 Java EE 的最新规范。

对于业务逻辑层，组件是 JavaBean 和 EJB。JavaBean 运行在 Web 容器，用于实现业务逻辑，EJB 组件用于实现分布式的业务逻辑，它的运行需要专门的 EJB 容器。Tomcat 服务器不是 EJB 容器，不能运行 EJB。常用的 EJB 容器是 JBoss，它是开源的，早期的版本只支持 EJB，不支持 JSP、Servlet 组件的运行，后期推出的 JBoss 作为一个完整的应用服务器，既是一个 Web 容器，也是一个 EJB 容器。

组件在容器中运行，还需要有提供通信服务的技术和其他提供基础服务的技术作为支持。通信技术包括 Internet 协议、RMI 远程方法调用、OMGP 对象管理组协议、消息技术等；服务技术有命名服务、数据库连接、事务处理、安全服务等。

1.3 Java EE 的主要技术

Java EE 的技术规范有很多，每一种技术都是以 API 的形式提供给编程人员，并且有相应的编写规范。这些 API 在 Java EE 的 SDK 软件开发工具中集中提供。当用户在开发平台中选定一个 Java EE 版本时，开发平台提供相应的 SDK。

1.3.1 Servlet 技术

Servlet 技术是用于动态处理客户端请求并生成页面的一种技术，实现服务器的功能。Servlet 程序是运行在服务器端的 Java 程序，它的编写要遵循 Servlet 规范。最初，网页是由 Servlet 生成的，由于利用该技术生成网页比较烦琐，随后才推出 JSP 技术，专门用于生成动态网页。

1.3.2　JSP 技术

JSP（Java Server Page）即 Java 服务器端的页面，是 Sun 公司推出的一种动态网页技术标准，用于生成动态网页，类似于 ASP、PHP 等技术，是在静态网页 HTML 的基础上加上服务器端的代码，以生成动态的内容。JSP 是在 Servlet 的基础上开发的，运行时在 Web 容器内被自动转换成 Servlet 程序。

1.3.3　JavaBean 技术

JavaBean 技术是 Java 平台下的一种组件模型，它是一种软件代码复用技术，通过将业务逻辑封装成 JavaBean，使得代码复用，系统结构更加清晰。

1.3.4　JDBC 技术

JDBC（Java Database Connectivity）技术，即 Java 数据库连接技术，通过提供 Java 语言编写的类和接口实现对各种关系型数据库的连接，以统一的方式操作数据库。

1.3.5　JSF 技术

JSF（Java Server Faces）是一个用来创建 Web 应用程序的用户界面框架，是 Java EE 官方提供的实现 MVC 模式的框架，采用了组件模型和事件驱动，提高了用户创建应用程序的效率，也有利于维护。JSF 所采用的组件框架包括很多功能，如输入校验、事件处理、模型对象和组件之间的数据转换、创建可管理的模型对象、页面导航配置和表达式语言，极大地简化了 Web 应用程序开发的复杂性。

1.3.6　JNDI 技术

JNDI（Java Name and Directory Interface）是 Java 命名和目录接口。JNDI API 提供了命名和目录功能，使得应用程序可以访问多种命名和目录服务。在 Java EE 中，有多种组件，如应用程序客户端、EJB、Web 组件和数据源等，这些组件可能分布在不同的服务器上，JNDI 规范各种组件的命名，如表示 EJB 组件，采用 Java:comp/env/ejb 来表示，JDBC 数据源则采用 Java:comp/env/jdbc，并提供一组方法对各种组件进行访问。

1.3.7　EJB 技术

EJB（Enterprise JavaBean）是企业级 Java 组件，是一种 Java 服务器端组件模型，专为分布式应用开发而设计，在复杂多变的企业级应用中，分布在不同部门、机构的业务逻辑可以编写成 EJB，部署在不同的服务器，使用 JNDI 规范这些组件的命名，进而对这些组件进行访问，从而使得分布在不同位置的组件能够整合在一个应用系统中。早期的 EJB 是属于重量级的，使用非常不方便，需要进行大量的配置，创建并实现大量的接口，到了 Java EE 5 的 EJB 3.0 时，EJB 借鉴流行的轻量级框架技术，进行了大量的改进，降低了其开发难度，使其变成轻量级。尽管如此，随着流行的开源框架如 SSH、SSM 被开发人员所接受，EJB 还是没有被广泛地应用。

1.3.8　JMS 技术

JMS(Java Message Service)是 Java 的消息服务,是用于和面向消息的中间件相互通信的应用程序接口,提供创建、发送、接收和读取消息的服务,既支持点对点的消息模型,也支持发布/订阅的消息模型。

1.3.9　JavaMail 技术

JavaMail 提供一组用于实现收发电子邮件的应用程序编程接口,可以在企业级应用系统中实现收发电子邮件。使用 James 搭建邮件服务器,然后使用 JavaMail API 实现 James 邮件服务器邮件的收发。

1.3.10　JPA 技术

JPA(Java Persistence API,Java 持久化 API)是一个基于 Java 标准的持久化解决方案,通过对象/关系映射的方式,弥补了面向对象模型和关系型数据库之间的差异。JPA 包含查询语言、对象/关系映射元数据。

1.3.11　JTA 技术

在应用系统中,相互依赖的数据库访问操作可设置为一个事务,JTA(Java Transaction API,Java 事务 API)提供了对事务处理的接口,实现事务提交和回滚,以确保事务的原子性、一致性、隔离性和持久性,默认是自动提交。

1.3.12　Web 服务技术

Web 服务技术是基于 HTTP 协议的一种跨平台、跨语言的可互操作的服务。在分布式环境下,各个系统所采用的平台和编程语言不可能是完全相同的,在这种异构环境下,借助 Web 服务技术,将需要互操作的业务编写成 Web 服务,然后使用 WSDL(Web Services Description Language,Web Service 描述语言)进行服务的发布,再以 SOAP(Simple Object Access Protocal)即简单对象访问协议的形式通过 HTTP 进行传输,实现了客户端请求和 Web 服务响应,以实现分布式环境下业务的集成。

Java EE 所提供的 Web 服务支持包括:

- JAX-WS 2.2(Java API for XML Web Services):Java EE 平台用于开发 Web 服务的 API,定义了访问 Web 服务的客户端 API 和实现 Web Service endpoint 的技术。
- JAXB 2.2(Java Architecture for XML Binding):Java XML 绑定架构,使用 JAXB 将 XML 数据转换为 Java 对象。
- SAAJ(SOAP with Attachments API for Java):JAX-WS 依赖的底层 API,SAAJ 产生和接收 SOA 1.1、SOA 1.2 规范,以及 SOAP with Attachments 记录的消息。

1.3.13 JAAS 技术

JAAS(Java Authentication and Authorization Service)技术是指 Java 认证和授权服务,支持对运行在 Java EE 应用程序上的某个指定用户或用户组进行认证和授权。通过 JAAS,编程人员能够将一些标准的安全机制通过一种通用的、可配置的方式集成到系统中。

1.4 Java EE 的开发环境

1.4.1 本书使用的开发环境

- 操作系统:Windows 10,64 位。
- JDK:jdk12.0.2。安装文件:jdk-12.0.2_windows-x64_bin.exe。
- IDE:Eclipse 2021.3。安装文件:eclipse-jee-2021-03-R-win32-x86_64.zip。
- Web 服务器:Tomcate 9.0.4。安装文件:Aapache-tomcat-9.0.45-windows-x64.zip。
- 数据库系统:MySQL 8.0.24。安装文件:mysql-installer-community-8.0.24.0.msi。

1.4.2 JDK 安装、配置和测试

1. 安装

JDK(Java Development Kit,Java 开发工具包)是 Java EE 平台应用程序的基础,包括 Java 运行环境、Java 工具和 Java 基础类库。Java 最初是由 Sun 公司发布的,Sun 公司后来被 Oracle 公司收购,下载 JDK 的官网是:https://www.oracle.com/technetwork/Java/Javase/downloads/index.html。

下载 64 位 Windows 平台下的安装文件 jdk-12.0.2_windows-x64_bin.exe,双击进行安装,安装目录默认为 C:\Program Files\Java\jdk-12.0.2,过程如图 1.4 和图 1.5 所示。

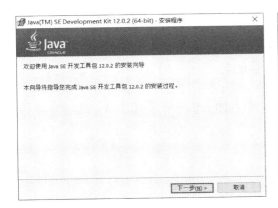

图 1.4 JDK 安装过程

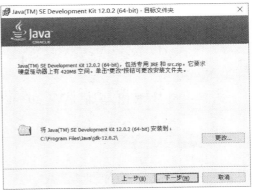

图 1.5 设置 JDK 安装目录

安装完成后,打开 C:\Program Files\Java 目录,可以看到 Java 目录中增加了一个

jdk-12.0.2 目录,如图 1.6 所示。

2. 配置

接下来,需要设置 JAVA_HOME、PATH 和 CLASSPATH 三个环境变量,通过"我的电脑→计算机→属性→高级系统设置",在"系统属性"对话框中进行设置,如图 1.7 所示。

图 1.6 Java 目录的内容

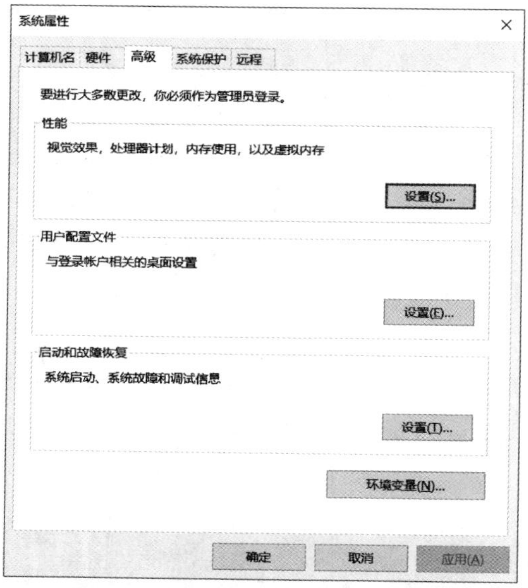

图 1.7 系统属性

点击"环境变量"按钮,添加"JAVA_HOME"变量,如图 1.8 所示,指明 Java 的安装目录。

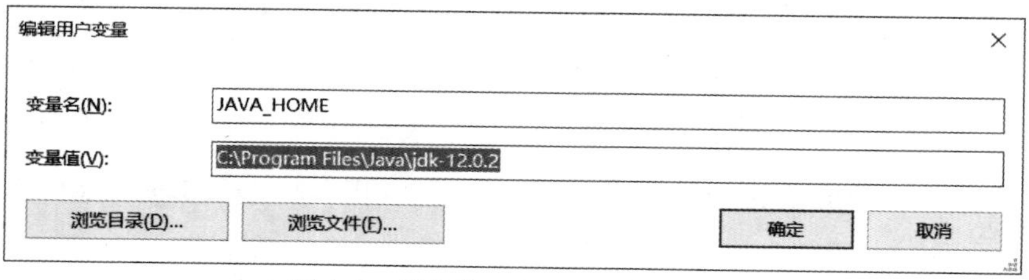

图 1.8 编辑"JAVA_HOME"变量

接着再编辑 PATH 环境变量,对于 PATH 变量,各种软件都需要设置,所以这里是在现有的基础上进行编辑的,设置 Java 的 PATH 值:"%JAVA_HOME%\bin;",如图1.9所示。

图1.9　编辑 PATH 变量

再添加"CLASSPATH"变量,变量值设置为"%JAVA_HOME%\lib\jrt-fs.jar;",这里指明搜索类的路径,如图1.10所示,多项之间用分号隔开。

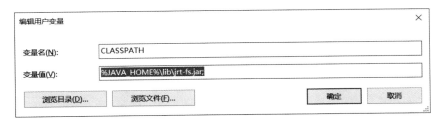

图1.10　编辑 CLASSPATH 变量

3. 测试

JDK 安装配置完成后,通过 cmd 打开命令窗口,输入"java-version",能够正常显示安装的 Java 的版本号,则说明 Java 的安装和配置正确,如图1.11所示。

1.4.3　Tomcat 服务器的安装和配置

服务器采用 Tomcat,它是一个免费开源的 Web 服务器,是一个高效、轻便的 JSP 和 Servlet 容器,被广泛地应用于中小型网站。Tomcat 可以从官网 http://tomcat.apache.org 上下载,在下载页面上选择"Core 中的 64-位 windows zip(pgp,sha512)"这一项,如图1.12所示,双击下载,得到 apache-tomcat-9.0.45-windows-x64.zip 压缩包。

双击压缩包进行解压,解压到 C:\Program Files\apache-tomcat-9.0.45 目录,目录结构如

图 1.13 所示。

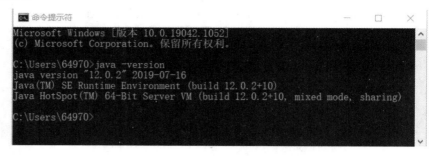

图 1.11 显示 Java 版本号

图 1.12 Tomcat 下载页面

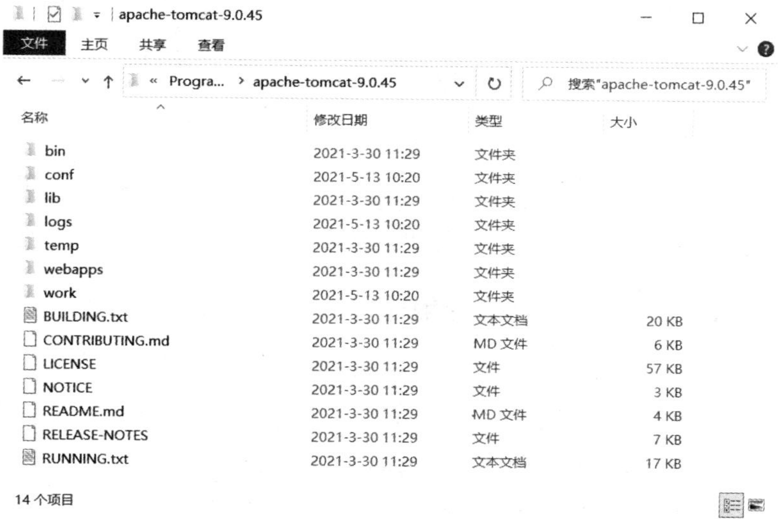

图 1.13 Tomcat 的目录结构

安装完成后,要配置环境变量,即通过"我的电脑→计算机→属性→高级系统设置",在"系统属性"对话框中进行设置,点击"环境变量"按钮,添加"CATALINA_HOME"变量,指明Tomcat的安装目录,如图1.14所示。

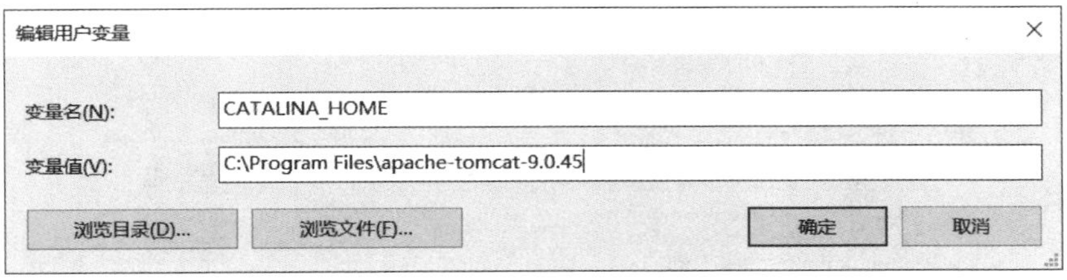

图1.14 设置 CATALINA_HOME 环境变量

接着,再增加"PATH"环境变量的设置值,添加"% CATALINA_HOME% \bin;""% CATALINA_HOME% \lib;""% CATALINA_HOME% \lib \servlet-api.jar;""% CATALINA_HOME% \lib\jsp-api.jar;",如图1.15所示。

图1.15 PATH 变量的设置值

最后,进入 Tomcat 安装目录中的 bin 目录,运行 startup.bat,启动服务器,打开浏览器,输入 localhost:8080,如果看到图1.16所示的页面,说明服务器安装成功。

1.4.4 Eclipse IDE 安装和配置

1. 安装

Java EE 开发平台有很多种,常用的有 Eclipse 和 MyEclipse。Eclipse 是 IBM 推出的开源开发平台,支持包括 Java 在内的多种语言的开发,采用插件的机制,是一种可扩展、可配置的集成开发环境。Eclipse 以其完全开放的架构和强大的功能得到了几乎所有 Java 厂商的支持,并获得大部分开发人员的青睐。Eclipse 可以在 http://eclipse.org/downloads 官网上下载,如图 1.17 所示,选择"Eclipse IDE for Enterprise Java and Web Developers"版本,支持 Java EE 开发。本书采用的是 Eclipse 2021.3 版本,下载的文件是 eclipse-jee-2021-03-R-win32-x86_64.zip。

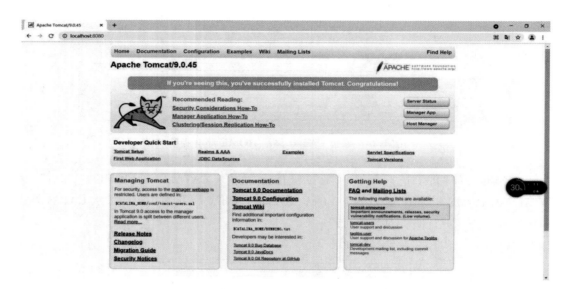

图 1.16　Tomcat 服务器启动成功的页面

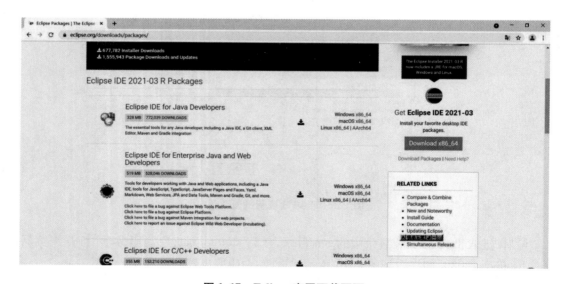

图 1.17　Eclipse 官网下载页面

对压缩文件进行解压,双击 eclipse.exe 就可以运行。第一次启动 Eclipse 时,为其指定工作空间 Workspace,我们创建的项目就保存在工作空间中,如图 1.18、图 1.19 所示。

2. 配置

接下来进行环境配置。首先,设置 JRE ,在 Eclipse 中选择"window→preferences"菜单项，打开"Preferences"设置对话框，如图 1.20 所示，展开 Java 项，在"Installed JREs"中添加安装的 JDK，如图 1.21 所示。

图 1.18　Eclipse 的目录

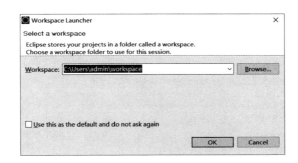

图 1.19　设置 Eclipse 的工作空间

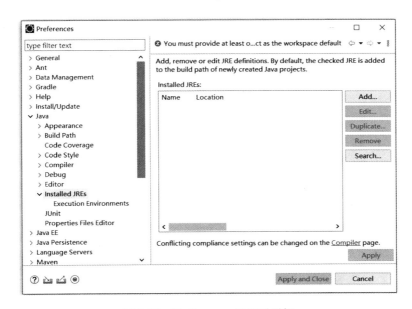

图 1.20　Preferences 设置对话框

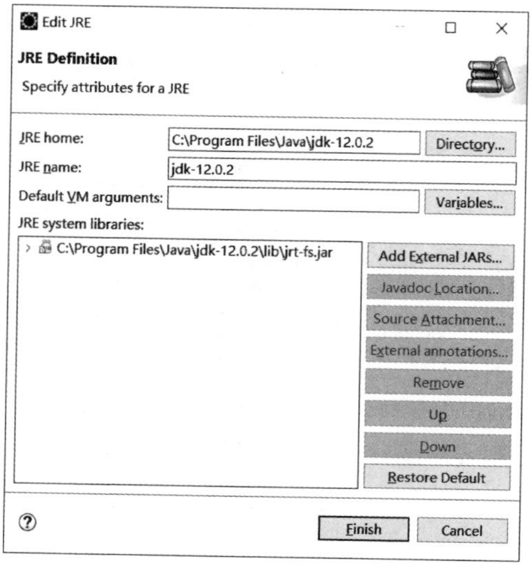

图 1.21 配置 JRE

接下来,配置 Tomcat 服务器,在"Preferences"对话框中,展开"Server"项,选中"Runtime Environment"项,点击"Add"按钮,添加所安装的 Tomcat 服务器,如图 1.22 所示。

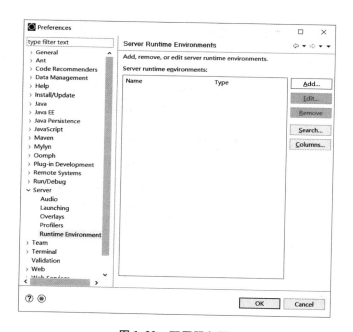

图 1.22 配置服务器

选中"Apache Tomcat v9.0",然后指定 Tomcat 的安装目录和 JRE,如图 1.23 所示。

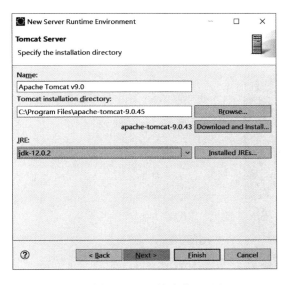

图 1.23　选择 Tomcat 的安装目录和 JRE

1.4.5　MySQL 数据库的安装

在 Java EE 项目开发中，需要使用数据库来保存数据，通常采用 MySQL 数据库。MySQL 是一款免费、开源的中小型关系型数据库管理系统，由于其体积小、速度快、免费、开源，目前被广泛地应用在中小型网站中。其官网下载地址是：http://dev.mysql.com/downloads/mysql，页面如图 1.24 所示，下载 MySQL Community Server 8.0.24。

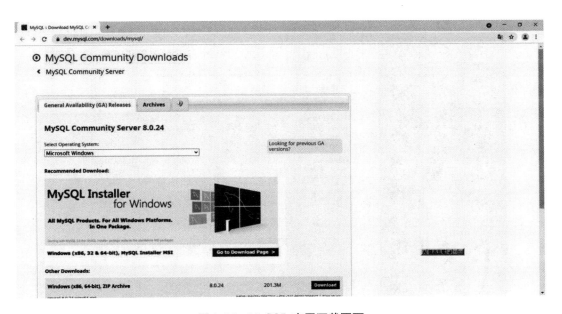

图 1.24　MySQL 官网下载页面

双击下载的 mysql-installer-community-8.0.24.msi 文件进行安装，如图 1.25 所示，选择

安装类型为"Custom",进行自定义安装。

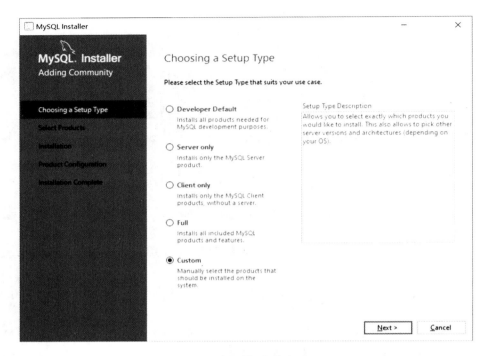

图 1.25　选择安装类型

如图 1.26 所示,选择安装 MySQL Server 8.0。

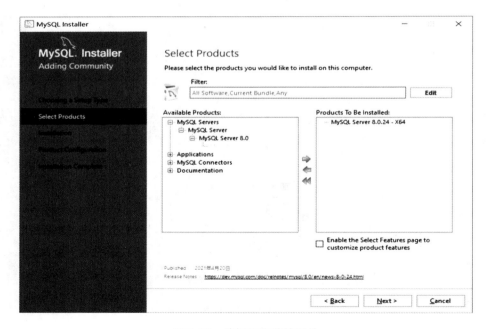

图 1.26　选择要安装的组件

安装完成后进行配置，按照图 1.27 所示配置类型和网络连接，其中 TCP/IP 连接的端口是 3306。

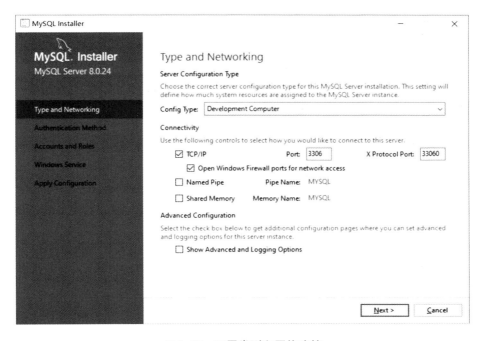

图 1.27　配置类型和网络连接

接下来设置 root 用户名和密码，如图 1.28 所示。

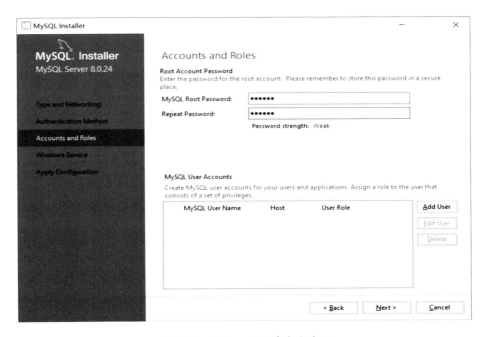

图 1.28　设置 root 用户名和密码

完成安装后,可通过命令行的方式访问 MySQL,输入 root 的密码,如图 1.29 所示,按回车,可以显示当前 MySQL 的版本是 8.0.24,表示 MySQL 已成功启动。

图 1.29　命令行方式访问 MySQL

习　　题

1. Java EE 常用技术有哪些?

2. 请画出企业级应用软件的 4 层体系结构,并介绍 Java EE 对各层提供了哪些技术支持?

3. 请按照课本的步骤,搭建 Java EE 的开发环境。

第 2 章　前端开发技术

【本章内容】

- 2.1　HTML 5 + DIV + CSS
- 2.2　JavaScript 技术
- 2.3　jQuery 框架
- 2.4　Bootstrap 框架

2.1　HTML 5 + DIV + CSS

2.1.1　HTML 5

HTML(Hyper Text Mark-up Language,超文本标记语言)是一种用于创建网页的标准标记语言,使用标记的形式来描述网页的结构和内容。

HTML 最早由欧洲原子核研究委员会的 Berners-Lee 发明,随后由 W3C 组织(World Wide Web Consortium)负责开发和制定。1996 年 4 月发布了 HTML 3.2,1997 年 12 月发布了 HTML 4,1999 年 12 月发布了 HTML 4.01。2012 年,HTML 5 定稿并逐渐被各种浏览器和开发平台所支持。作为新一代的 Web 技术领航者,HTML 5 着力于解决各浏览器之间的兼容问题和增强 Web 开发的能力,服务于 PC 端和移动端的应用。HTML 5 具有良好的语义特性,在网页的结构定义上,使用了很多具有语义的元素,如 section、article、nav、aside 等,使得文档结构更加清晰、明确,更好地配合 CSS 进行各部分样式的定义。

1. HTML 文档的编写和运行

HTML 文档是纯文本文档,可以用任意一款文本编辑器编写,保存后缀为 html 或 htm。一个简单的 HTML 文档的基本结构如图 2.1 所示。

第一行是文档类型声明,当前的声明表示是 HTML 5 的文档,接下来是以 < html > 开始,以 </html > 结束,包括头部 < head > </head > 和主体 < body > </body > ,头部通常包括标题 < title > </title > ,还有一些属性的设置、链接样式表 CSS 的定义、引用样式表 CSS 文件和 JavaScript 脚本。主体部分是网页内容部分,通过标签描述各种对象和通过标签属性或 CSS 进行对象的设置。

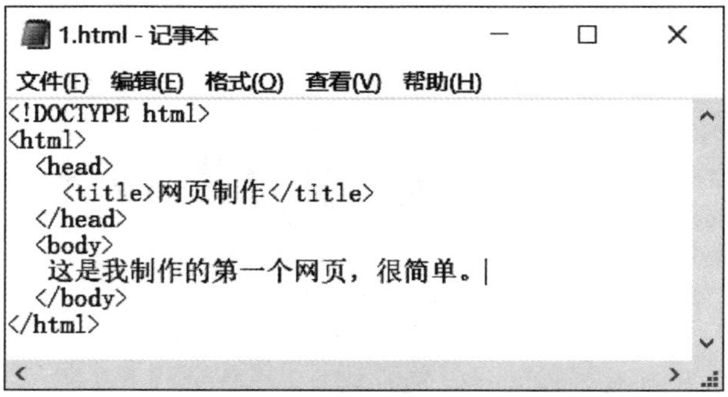

图 2.1　一个简单的 HTML 文档的基本结构

HTML 文档的运行是通过浏览器解析运行的,浏览器识别标签,标签定义了网页的结构和网页的内容。双击创建的 HTML 文档,可以在浏览器上显示网页的内容。上面创建的 1.html 文档,它的运行结果如图 2.2 所示。

图 2.2　HTML 文档的运行结果

2. HTML 的常用标签

HTML 的标签也称为标记,它是由一对尖括号括起来的,里面是英文字母或英文字母加上数字。标签通常是成对的,也有一些是单个的,如 < img > 和 < hr >。标签不区分大小写,但书写时统一大小写会比较规范,可读性也比较好。如果标签写错了,则运行时是看不到想要的效果的。浏览器只识别标签,通过标签正确显示网页的内容。标签有很多,下面介绍常用的几种标签。

（1）段落标签、换行标签和水平线标签

< p > </p>:段落标签,段与段之间会空一行。

< br >:换行标签,是单标签,标签后面的内容另起一行。

< hr >:水平线标签,也是单标签,用于插入一条水平线。水平线标签包含几个常用的属性:width 属性用于设置水平线的宽度,可以用绝对值,表示宽度是多少像素,也可以用相对值,表示宽度占当前浏览器窗口的百分比;size 属性用于设置水平线的粗细,单位是像素;color 属性用于设置水平线的颜色。

段落标签、换行标签和水平线标签示例代码如图 2.3 所示,运行结果如图 2.4 所示。

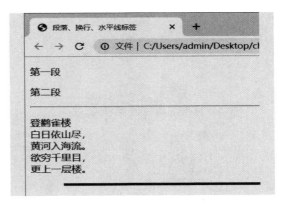

图2.3　段落标签、换行标签和水平线标签示例代码　　图2.4　段落标签、换行标签和水平线标签示例代码运行结果

（2）标题标签

有6级标题标签，即从<h1>到<h6>，作为标题，它的格式是加粗并居中，独占一行，各级标题的字体大小不一样，h1字体最大，h6字体最小。标题标签示例代码如图2.5所示，运行结果如图2.6所示。

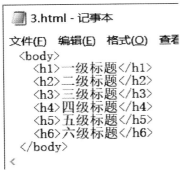

图2.5　标题标签示例代码　　图2.6　标题标签示例代码运行结果

（3）列表标签

<ol></ol>：有序列表标签。

<ul></ul>：无序列表标签。

<li></li>：列表项标签。

列表标签可以有多重嵌套，可以用来制作导航栏。列表标签示例代码如图2.7所示，运行结果如图2.8所示。

（4）图片标签

图片标签<img>是单标签，用于在网页上显示图片。图片标签包含几个常用的属性：src属性是必需的属性，用于指定要显示的图片，包括目录和文件名，如果只是文件名，表示图片跟网页文档处于同一个目录；width和height属性是可选属性，用于设置图片的宽度和高度，可用像素表示，也可用百分比表示；alt属性是可选属性，用来进行文字提示，当图片没办法加载时,在插入图片的位置就可以看到相应的文字提示。图片标签示例代码如图 2.9 所示,运行结果如图 2.10 所示。

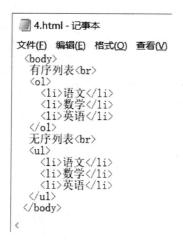

图 2.7　列表标签示例代码

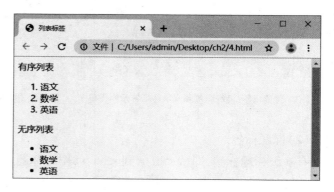

图 2.8　列表标签示例代码运行结果

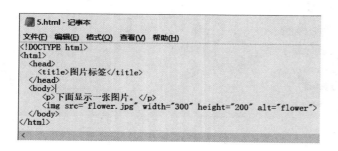

图 2.9　图片标签示例代码

图 2.10　图片标签示例代码运行结果

(5)表格标签

<table> </table>:表格的标签,width 和 height 属性用于设置表格的宽度和高度,cellspacing 属性用于设置单元格间距,cellpadding 用于设置单元格的填充距离,bgcolor 属性用于设置背景色。<tr> 和 <td> 标签也包括 bgcolor 属性。

<tr> </tr>:行的标签。

<td> </td>:单元格的标签,colspan 属性设置横向合并单元格,rowspan 属性设置纵向合并单元格。

表格标签示例代码如图 2.11 所示,运行结果如图 2.12 所示。

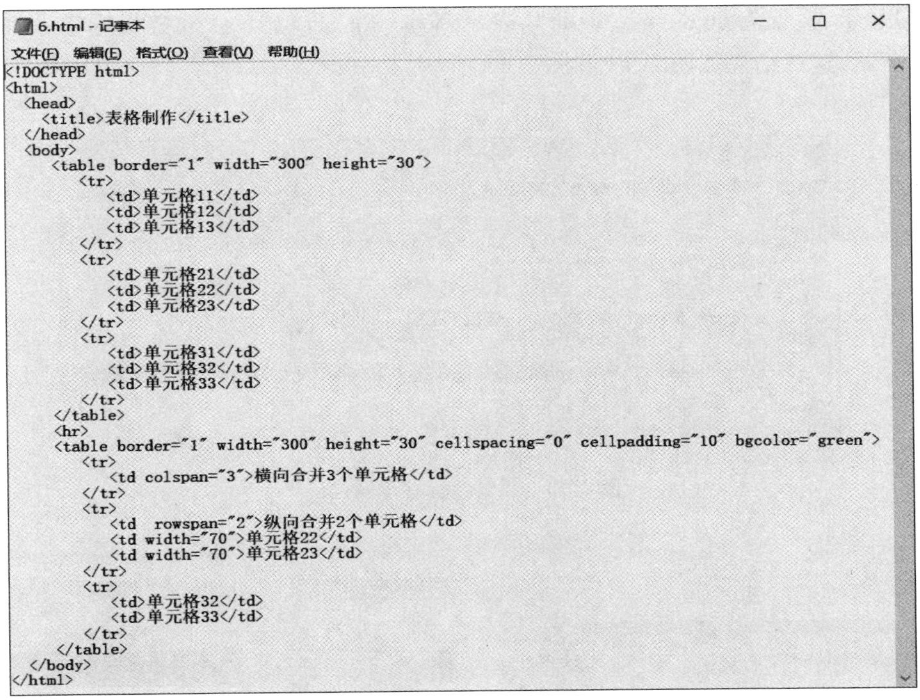


```
<!DOCTYPE html>
<html>
  <head>
    <title>表格制作</title>
  </head>
  <body>
     <table border="1" width="300" height="30">
        <tr>
           <td>单元格11</td>
           <td>单元格12</td>
           <td>单元格13</td>
        </tr>
        <tr>
           <td>单元格21</td>
           <td>单元格22</td>
           <td>单元格23</td>
        </tr>
        <tr>
           <td>单元格31</td>
           <td>单元格32</td>
           <td>单元格33</td>
        </tr>
     </table>
     <hr>
     <table border="1" width="300" height="30" cellspacing="0" cellpadding="10" bgcolor="green">
        <tr>
           <td colspan="3">横向合并3个单元格</td>
        </tr>
        <tr>
           <td  rowspan="2">纵向合并2个单元格</td>
           <td width="70">单元格22</td>
           <td width="70">单元格23</td>
        </tr>
        <tr>
           <td>单元格32</td>
           <td>单元格33</td>
        </tr>
     </table>
  </body>
</html>
```

图 2.11　表格标签示例代码

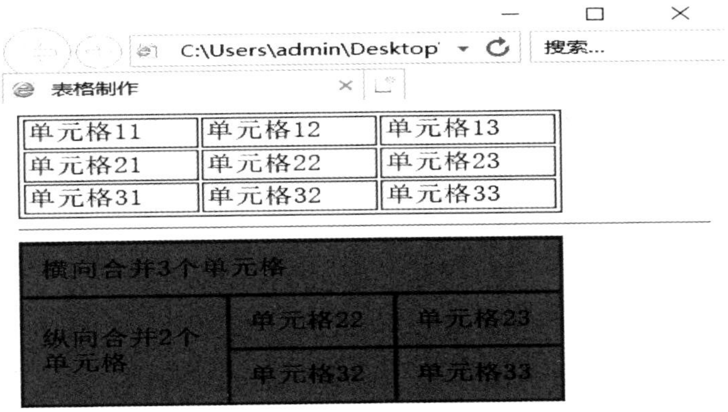

图 2.12　表格标签示例代码运行结果

(6)超链接标签

<a> </a>:超链接标签,href 属性是必需的属性,用于指明链接的目标,可以是另一个网页或其他文件,也可以是外网,但链接到外网时,网址前面必须加上访问协议如"http://",不然所设置的网址就会被当成目录。指明文件时,路径的表示要正确,"./"表示当前目录,".."表示上一级目录。如果没有路径,直接是文件名,表示要打开的文件与当前网页处于同一个目录。

target 属性是可选属性,用于指明是在当前窗口显示还是新建一个窗口显示,默认是在

当前窗口显示,设置为 target = "_blank",表示新建一个窗口显示。超链接标签示例代码如图 2.13 所示,运行结果如图 2.14 和图 2.15 所示。

图 2.13　超链接标签示例代码

图 2.14　超链接标签示例代码运行结果 1

图 2.15　超链接标签示例代码运行结果 2

(7)表单标签

表单是动态网页技术的一个重要部分,通过表单可以将用户输入的数据提交到服务器,由指定的程序处理,实现客户端与服务器端的交互。HTML 5 相比 HTML 4,提供了更多的表单元素,方便输入各种数据类型。

< form > </ form > :表单标签。name 属性用于设置表单名称;method 属性用于设置提交方式,有 get 和 post 两个值;action 属性指定处理表单的程序。

< input > :输入元素。输入元素包含各种类型,可通过 type 属性设置。在 HTML 5 中,新增了很多的输入类型,并且功能增强了很多。input 类型和用法如表 2.1 所示。

表 2.1　input 类型和用法

类型名	作用	说明
text	文本框	< input type = "text" name = "username" value = "liming" placeholder = "输入用户名"/> HTML 5 新增 placeholder 属性,用于设置提示文字
password	密码框	< input type = "password" name = "pwd" />

表 2.1(续)

类型名	作用	说明
radio	单选按钮	< input type = " radio"　name = " sex"　　value = " male" >男 < input type = " radio"　name = " sex"　　value = " female" >女 同一组按钮,name 值要相同
checkbox	复选框	< input type = " checkbox"　name = " hobby"　value = " 看电影" >看电影 < input type = " checkbox"　name = " hobby"　　value = " 游泳" >游泳 < input type = " checkbox"　name = " hobby"　　value = " 旅游" >旅游 < input type = " checkbox"　name = " hobby"　　value = " 运动" >运动 同一组复选框,name 值要相同
submit	提交按钮	< input type = " submit"　value = " 登录"/ >
reset	重置按钮	< input type = " reset"　value = " 重置"/ >
button	普通按钮	< input type = " button"　value = " 注册"　onClick = " registCheck( )"/ > 设置为普通按钮,是为了将数据交给指定的 JavaScript 程序处理,通常用于数据的校验或进行信息的确认
hidden	隐藏域	< input type = " hidden"　value = " 12"/ >,隐藏域不会在页面上显示出来,但实际是存在的,通过这种方式可以传递一些数据
number	数值输入域	< input type = " number"　name = " age"　min = " 6"　max = " 14"　value = " 10"　step = " 2"/ >,min 和 max 设置最小值和最大值,value 设置默认值,step 设置步长
date	年、月、日	< input type = " date"　name = " birthDate"/ >,也可以是 month、week、time 类型,分别用于设置月、周和时间
datetime	时间	< input type = " datetime-local"　name = " publicDate"/ > datetime-local 设置本地时间,datetime 设置世界标准时间
url	输入 url	< input type = " url"　name = " homepage"/ > 要求输入的数据要符合 URL 格式
email	输入 email	< input type = " email"　name = " email"/ >,数据要符合 email 格式

列表框元素 < select > </select > :通过 < option > </option > 设置列表项,value 属性设置列表项的值,提示文字在 < option > </option > 标签中间。

文本域元素 < textarea > </textarea > :用于设置多行文本框,rows 属性设置行数,cols 属性设置列数。

图 2.16 是注册表单的代码,包括了各种类型的表单元素,可以参考其用法,运行结果如图 2.17 所示。

```
regist.html
<!DOCTYPE html>
<html>
  <head>
    <meta charset="utf-8">
    <title>注册表单</title>
  </head>
  <body>
     <form name="form1" method="post" action="doRegist.jsp">
    用户名：<input type="text" name="username"  placeholder="请输入用户名"/><br>
    密码：<input type="password" name="pwd"  placeholder="请输入密码"/><br>
  确认密码：<input type="password" name="rePwd"  placeholder="输入确认密码"/><br>
    性别：<input type="radio" name="sex"  value="male">男
              <input type="radio" name="sex"  value="female">女<br>
    所读专业：<select name="department">
                        <option value="数学">数学</option>
                        <option value="计算机">计算机</option>
                        <option value="英语">英语</option>
                        <option value="体育">体育</option>
                  </select><br>
    出生日期：<input type="date" name="birth"/><br>
        email地址：<input type="email" name="email"/><br>
    主页：<input type="url" name="homepage"/><br>
    兴趣爱好：<input type="checkbox" name="hobby" value="看电影">看电影
                  <input type="checkbox" name="hobby"  value="游泳">游泳
                  <input type="checkbox" name="hobby"  value="旅游">旅游
                  <input type="checkbox" name="hobby"  value="运动">运动<br>
    计划开始学习的时间：<input type="week" name="startWeek"/><br>
    备注：<textarea rows="5" cols="30"></textarea><br>
        <input type="submit" value="注册"/>  <input type="reset" value="重置"/> <br>
     </form>
  </body>
</html>
```

图 2.16　注册表单的代码

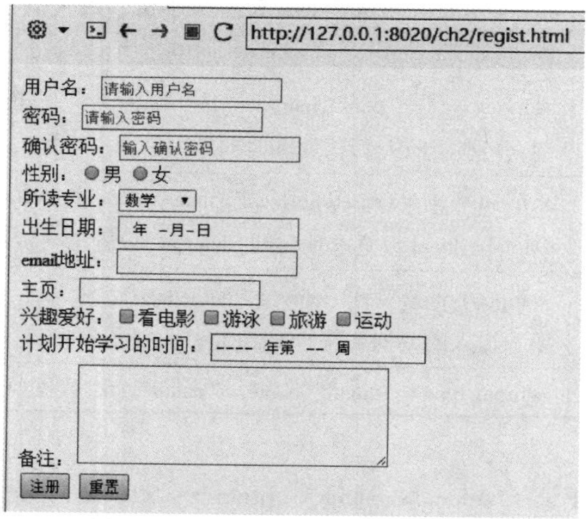

图 2.17　注册表单的代码运行结果

2.1.2　CSS

CSS(Cascading Style Sheet,层叠样式表)使用标记的形式来定义网页的样式,相比较使用 HTML 标记的属性来定义样式,它提供了丰富的样式定义标记,能够实现精准的控制,并且允许网页内容与样式定义相分离。当今的主流是采用 CSS 来进行网页样式定义。

CSS 样式的定义是由样式规则组成的,一条样式规则包括两部分,即选择器和声明。选择器就是样式作用的对象,有标签选择器、类选择器、ID 选择器和复合内容选择器;声明是具体的样式定义,由属性名和属性值组成。

CSS 样式应用有四种方式,分别是行内样式、嵌入式、链接式和导入式,其中嵌入式和链接式是常用的样式应用方式。嵌入式是将样式定义的代码放在网页的头部,通过 < style type = "text/css" > 和 </style > 标签引出,只作用于当前网页;链接式是采用独立文件的形式定义样式,所定义的样式可以应用到多个网页,从而使网站具有统一的风格。在图 2.18 所示的示例文件中,使用的是嵌入式样式表,在 < style type = "text/css" > 和 </style > 标签中,设置了两条样式规则,第一条设置选择器是 p 标签,重新定义了标签 p 的样式,设置其字体大小是 15 px,颜色是蓝色,背景颜色是黄色;第二条设置选择器是 img 标签,重新定义了图片的宽度是 300 px,高度是 200 px,边框是点线、绿色、5 px。CSS 示例文件的运行效果如图 2.19 所示。

图 2.18　CSS 示例文件　　　　图 2.19　CSS 示例文件的运行效果

CSS 示例文件中的样式定义,也可以改成定义独立的样式表文件"sample.css"。样式表文件包含多条样式规则,如图 2.20 所示。链接式样式表文件应用于两个网页的效果,如图 2.21 所示。

在样式的定义中,使用标签选择器不太灵活,当设置一个标签的样式时,网页中所有这类标签都被应用新的样式。同时,有的时候需要一种样式应用于几种标签,在这种情况下,可以使用类选择器。类选择器就是将一种样式定义为类,类名由用户命名,不能与标签名相同,类名前面要加上"."。应用样式时,哪个标签需要应用这种样式,就在这个标签中通过"class"属性名来指定。类选择器的示例代码和运行效果如图 2.22、图 2.23 所示。

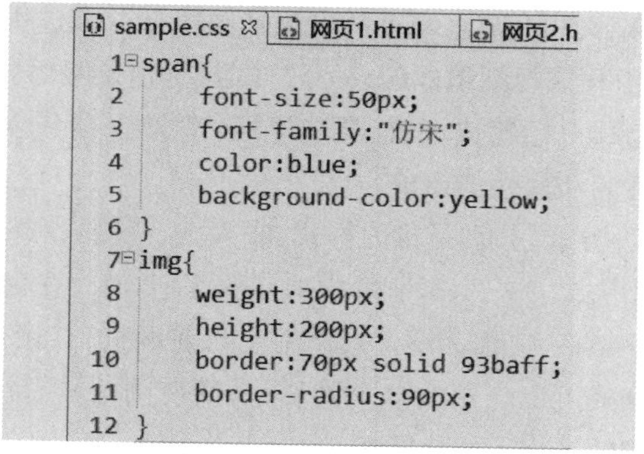

```
span{
    font-size:50px;
    font-family:"仿宋";
    color:blue;
    background-color:yellow;
}
img{
    weight:300px;
    height:200px;
    border:70px solid 93baff;
    border-radius:90px;
}
```

图 2.20　链接式样式表文件

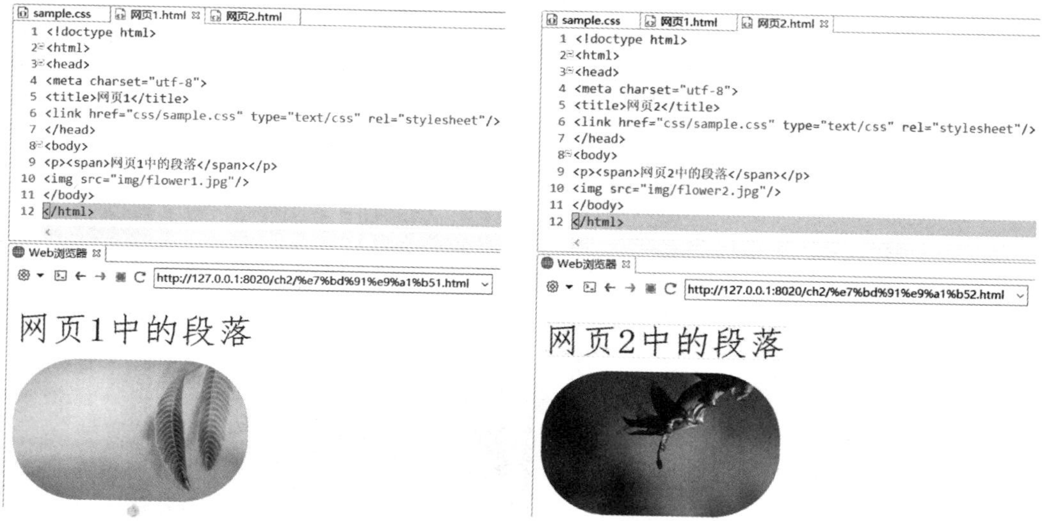

```
<!doctype html>
<html>
<head>
<meta charset="utf-8">
<title>网页1</title>
<link href="css/sample.css" type="text/css" rel="stylesheet"/>
</head>
<body>
<p><span>网页1中的段落</span></p>
<img src="img/flower1.jpg"/>
</body>
</html>
```

```
<!doctype html>
<html>
<head>
<meta charset="utf-8">
<title>网页2</title>
<link href="css/sample.css" type="text/css" rel="stylesheet"/>
</head>
<body>
<p><span>网页2中的段落</span></p>
<img src="img/flower2.jpg"/>
</body>
</html>
```


图 2.21　链接式样式表文件应用于两个网页的效果

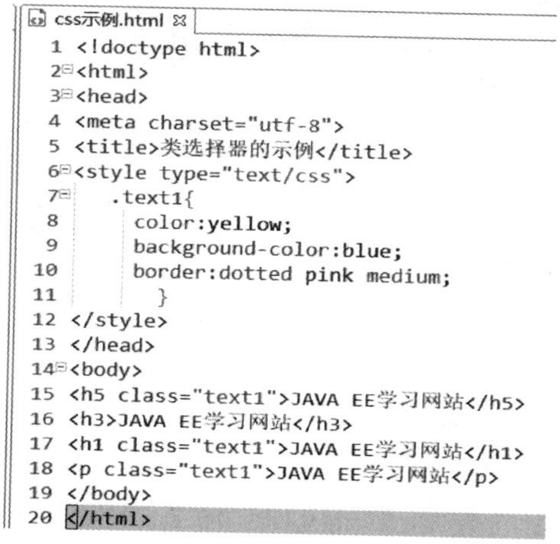

```
<!doctype html>
<html>
<head>
<meta charset="utf-8">
<title>类选择器的示例</title>
<style type="text/css">
    .text1{
      color:yellow;
      background-color:blue;
      border:dotted pink medium;
        }
</style>
</head>
<body>
<h5 class="text1">JAVA EE学习网站</h5>
<h3>JAVA EE学习网站</h3>
<h1 class="text1">JAVA EE学习网站</h1>
<p class="text1">JAVA EE学习网站</p>
</body>
</html>
```

图 2.22　类选择器的示例代码

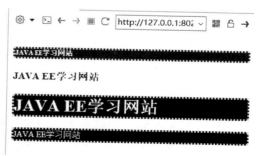

图 2.23　类选择器示例代码的运行效果

与类选择器相似的是 ID 选择器,定义时使用"#"号,应用时使用"id"作为属性名,属性值是唯一的。

复合选择器包括集体声明选择器、嵌套选择器等。集体声明选择器针对的是那些样式定义相同的选择器,定义时把这些选择器写在一起,中间用逗号隔开,同时进行声明。集体声明选择器的示例代码如图 2.24 所示,运行效果如图 2.25 所示。

图 2.24　集体声明选择器的示例代码　　图 2.25　集体声明选择器示例代码的运行效果

嵌套选择器指的是通过多层路径指定选择器,不是这一路径下的标签就不会应用这种样式。在指定路径时,中间用空格隔开。图 2.26 显示的是嵌套选择器的示例代码,图 2.27 显示的是运行效果。

在代码中,我们看到,网页中有两个 div 标签,第一个 div 标签用于显示导航栏,第二个 div 标签用于显示主要内容,两个 div 标签都是通过项目列表的形式来显示的。要设置导航栏的样式,这时要定位第一个 div 标签里面的 ul 标签里面的 li 标签,通过嵌套选择器的方式来定位;第二个 div 标签也是这样的结构,在 div 标签中要通过 id 来区分,所以选择器就表示为:div#nav ul li 或者#nav ul li。

还有常用的伪类选择器,通常是用在超链接的几种不同状态上。超链接有 4 种不同的状态:未访问链接 link、已访问链接 visited、鼠标停留在链接上 hover 和激活链接 active。当要设置这几种状态的不同样式时,选择器就表示为:a:link、a:visited、a:hover 和 a:active。这种表示方式称为伪类选择器。

```
嵌套选择器.html
<!doctype html>
<html>
<head>
<meta charset="utf-8">
<title>嵌套选择器</title>
<style type="text/css">
  div#nav ul li{
    list-style:none;
    }
  #nav ul li a{
    font-size:12px;
    color:red;
    text-decoration:none;
    }
</style>
</head>
<body>
<div id="nav">
<ul>
  <li><a href="#">菜单1</a></li>
  <li><a href="#">菜单2</a></li>
  <li><a href="#">菜单3</a></li>
</ul>
</div>
<div id="content">
<ul>
  <li><a href="#">内容1</a></li>
  <li><a href="#">内容2</a></li>
  <li><a href="#">内容3</a></li>
</ul>
</div>
<p> </p>
</body>
</html>
```

图 2.26　嵌套选择器的示例代码

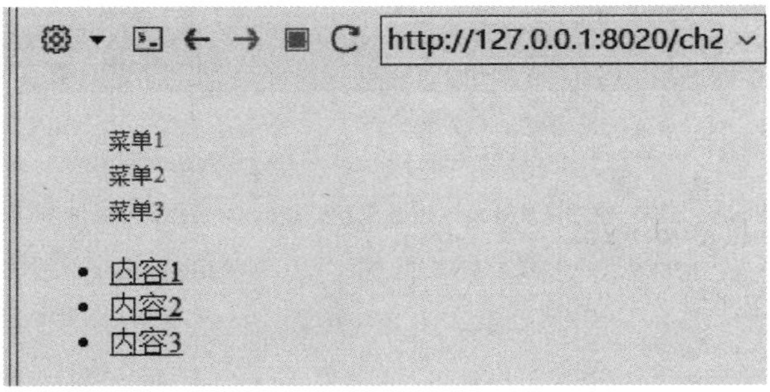

图 2.27　嵌套选择器示例代码的运行效果

2.1.3　HBuilder 开发工具

HBuilder 是 DCloud(数字天堂)推出的一款支持 HTML 5 的 Web 开发 IDE,是由 Java、C 和 Ruby 多种程序设计语言编写而成的,主体由 Java 编写,基于 Eclipse,兼容了 Eclipse 的插件。HBuilder 跟其他的 HTML 编辑器相比较,最大的特点就是“快”,HBuilder 使用代码块、快捷键、完整的语法提示和代码输入法等,大幅度提高 HTML、CSS 和 JS 文件的开发效率,同时具有“边改边看模式”,能够实时观看编辑效果。

下载 HBuilder. 9. 1. 29. windows. zip 压缩包,解压后运行文件夹 HBuilder 中的 HBuilder. exe 程序,启动界面如图 2.28 所示。

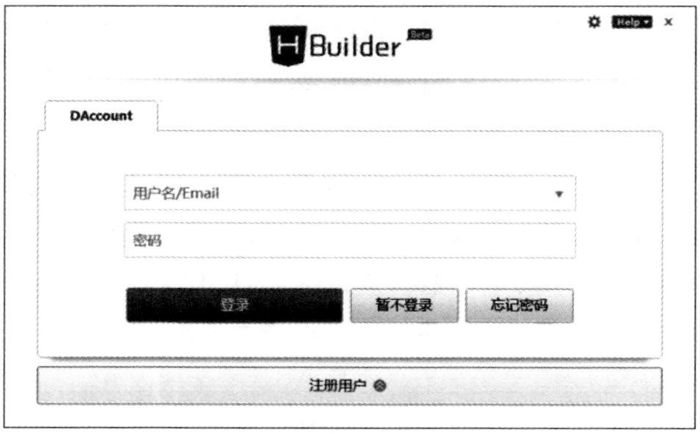

图 2.28　HBuilder 启动界面

选择"暂不登录"进入,启动后的软件界面如图 2.29 所示。

图 2.29 HBuilder 软件界面

通过文件下拉菜单选择"新建→Web 项目",打开"创建 Web 项目"对话框,输入项目名称"demo",选择"默认项目"模板,如图 2.30 所示,创建的"demo"项目的目录结构如图 2.31 所示,包含了 css、img 和 js 文件夹。

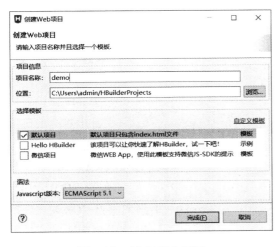

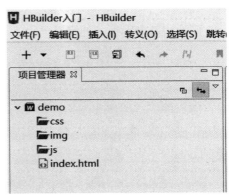

图 2.30 创建 Web 项目　　　　　　图 2.31 项目的目录结构

2.2 JavaScript 技术

2.2.1 JavaScript 简介

JavaScript 是一种客户端的脚本语言,编写的脚本代码嵌入在 HTML 代码中,通常是放在 <head> </head> 标签中,通过浏览器解释运行,实现客户端的动态效果。

在网页中插入 JavaScript 代码有两种方式。

一种方式是通过 <script language="JavaScript"></Script>方式,如图 2.32 所示。在网页中嵌入 JavaScript 代码,实现的效果是浏览网页时首先弹出一个对话框,如图 2.33 所示,当单击"OK"按钮后,可显示网页的内容。

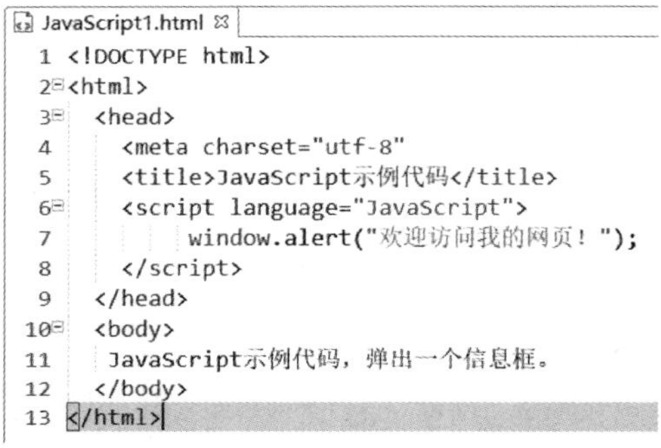

图 2.32 JavaScript 示例代码   图 2.33 JavaScript 示例代码的运行效果

另一种方式是将 JavaScript 代码独立为一个文件,即 js 文件。如图 2.34 所示,demo.js 就是一个脚本文件,然后在网页中加入一行代码 <script src="js/demo.js"></script>,如图 2.35 所示,从而引入 js 文件。

图 2.34 js 文件 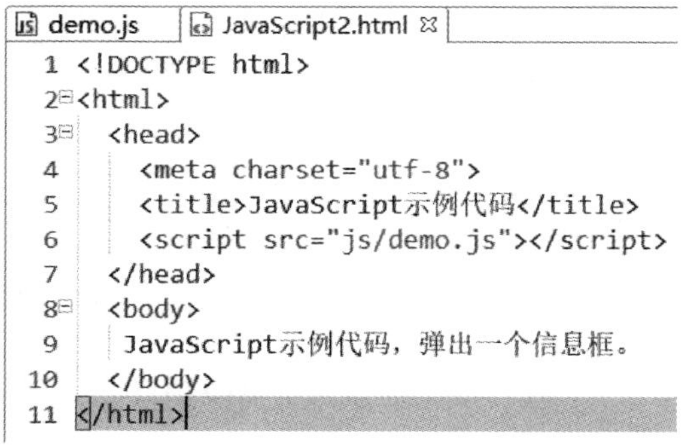   图 2.35 引入 js 文件的代码

2.2.2 JavaScript 基本语法

1. 变量的定义

JavaScript 语言定义变量时可以不指明类型,而是根据所赋值的类型来确定,即可以直接使用,不事先声明。语法如下:

```
var score;
```

定义变量 score,没有指明类型,也没有赋初值。

```
var a = 24,b = "good morning";
```

定义变量 a,并赋值 24,是整型类型;定义变量 b,并赋值 `"good morning"`,是字符串类型。

2. 基本语句

(1)条件语句

条件分支语句有 if 语句、if else 语句。

if 语句:

```
if(条件表达式)语句块
```

  if else 语句:

```
if(条件表达式)
    语句块1
else
    语句块2
```

  if else 语句的示例代码如图 2.36 所示,运行结果如图 2.37 所示。

```
JavaScript3.html
1  <!DOCTYPE html>
2  <html>
3    <head>
4      <title>if else语句</title>
5      <script language="JavaScript">
6        var score=75;
7        if (score>60)
8          document.write("及格");
9        else
10         document.write("不及格");
11
12     </script>
13   </head>
14   <body>
15   通过JavaScript代码给出评定
16   </body>
17  </html>
```

图 2.36　if else 语句的示例代码　　　　图 2.37　if else 语句示例代码的运行结果

从图 2.37 的运行结果看,JavaScript 代码放在 <head> </head> 之间,首先执行 <head> </head> 标签中的 JavaScript 代码,输出"及格"字符,然后输出 <body> </body> 标签中的内容"通过 JavaScript 代码给出评定"。

(2) 多分支语句

多分支语句可以采用 if… else if…else 语句或 switch 语句。if else 多分支语句的示例代码如图 2.38 所示，运行结果如图 2.39 所示。

```
<!DOCTYPE html>
<html>
  <head>
    <meta charset="utf-8">
    <title>if else多分支语句</title>
  </head>
  <body>
    通过JavaScript代码给出评定
    <script language="JavaScript">
      var score=75;
      document.write("你的分数是"+score+"<br>");
      if (score>90)
        document.write("优秀");
      else if(score>80)
        document.write("良好");
      else if(score>70)
        document.write("中等");
      else if(score>60)
        document.write("及格");
      else
        document.write("不及格");
    </script>
  </body>
</html>
```

图 2.38　if else 多分支语句的示例代码　　图 2.39　if else 多分支语句示例代码的运行结果

(3) for 语句

循环语句常用的是 for 语句。for 语句示例代码如图 2.40 所示，运行结果如图 2.41 所示。

```
<!DOCTYPE html>
<html>
  <head>
    <meta charset="utf-8">
    <title>for循环语句</title>
  </head>
  <body>
    计算1到5累加的和
    <script language="JavaScript">
      var sum=0;
      for(i=1;i<6;i++){
        sum=sum+i;
      }
      window.alert("1到5累加的和是："+sum);
    </script>
  </body>
</html>
```

图 2.40　for 语句示例代码　　图 2.41　for 语句示例代码的运行结果

3. 函数

在 JavaScript 中，函数既可以有返回值，也可以没有返回值，没有返回值相当于其他程序设计语言的过程，通过定义函数，可实现一定的功能，如进行数据的检验。图 2.42 所示的代码是通过定义函数输出一行“=”符号。函数示例代码的运行结果如图 2.43 所示。

```
<!DOCTYPE html>
<html>
  <head>
    <title>function函数</title>
    <script language="JavaScript">
     //编写一个函数,在页面中输出一行
        function check(){
       for(i=1;i<21;i++){
            document.write("=");
           }
         }
        check(); //调用函数
    </script>
  </head>
  <body>
  </body>
</html>
```

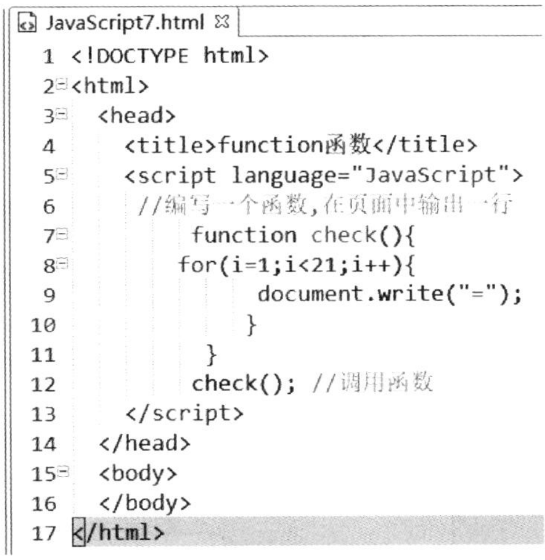

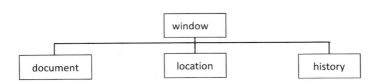

图 2.42 函数示例代码

图 2.43 函数示例代码的运行结果

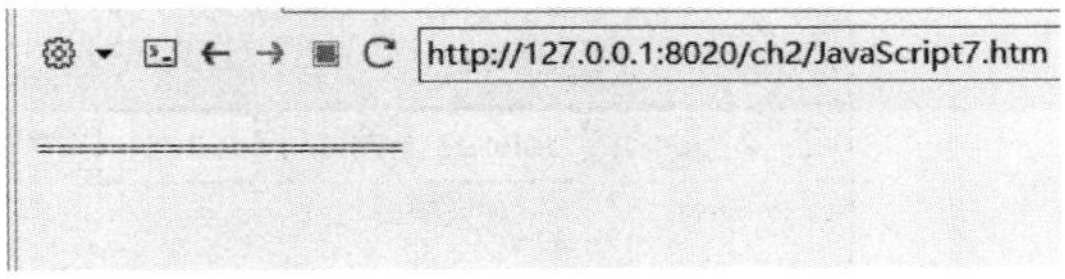

2.2.3 JavaScript 内置对象

JavaScript 采用面向对象的编程思想,提供了一系列的内置对象供用户编程,可实现对窗口等对象的操作。常用的内置对象及其从属关系如图 2.44 所示。

图 2.44 常用的内置对象及其从属关系

1. window 对象

window 对象常用的操作有弹出一个对话框和打开、关闭一个窗口,示例代码和运行效果如图 2.45、图 2.46 所示。

2. document 对象

document 对象用于操作网页,从属于 window 对象,常用的方法有 document. write( )方法,用于在网页上输出内容。另外通过 document 对象,可以获取网页元素的值,也常用于获取表单元素的值,然后对值进行判断。示例代码如图 2.47 所示,运行结果如图 2.48 所示。

document 对象的调用方法如下。

document. form1. username. value( ):获取表单 form1 中的文本框 username 的值。

document. form1. pwd. focus( ):密码框获得焦点。

document. location:得到当前网页的地址。

```
JavaScript8.html ⊠
 1  <!DOCTYPE html>
 2  <html>
 3      <head>
 4          <title>打开一个窗口</title>
 5          <script language="JavaScript">
 6              function openW(){
 7                  window.open("notice.html","高考喜讯",
 8                  "width=300,height=300,top=40,left=20,
 9                  toolbar=no,menubar=no,location=no,
10                  directories=no,scrollbars=no")
11              }
12          </script>
13      </head>
14      <body onload="openW()">
15      </body>
16  </html>
```

图 2.45　JavaScript 打开一个窗口的示例代码　　图 2.46　打开一个窗口示例代码的运行效果

```
JavaScript9.html ⊠
 1  <!DOCTYPE html>
 2  <html>
 3      <head>
 4          <meta charset="utf-8">
 5          <title>登录页面</title>
 6          <script language="JavaScript">
 7              function checkLogin(){
 8                  var username=document.form1.username.value.trim();
 9                  var password=document.form1.password.value.trim();
10                  if(username==""){
11                      alert("用户名不能为空");
12                      document.form1.username.focus()
13                      return;
14                  }
15                  if(password==""){
16                      alert("密码不能为空");
17                      document.form1.password.focus();
18                      return;
19                  }
20                  document.form1.submit();
21              }
22          </script>
23      </head>
24      <body>
25          <form name="form1" action="doLogin.jsp" method="post">
26              username:<input type="text" name="username"><br>
27              password:<input type="password" name="password"><br>
28              <input type="button" value="登录" onclick=checkLogin()><br>
29          </form>
30      </body>
31  </html>
```

图 2.47　JavaScript 实现表单数据检验示例代码

图 2.48　JavaScript 实现表单数据检验示例代码的运行结果

3. history 对象

history 对象保存了用户的历史操作记录,通过 history.back()和 history.forward()可以实现前进和后退。

4. location 对象

location 对象可以访问浏览器地址栏,常用 location.href 实现网页跳转。

2.3 jQuery 框架

2.3.1 jQuery 概述

jQuery 是继 prototype 之后的一个优秀的轻量级 JavaScript 框架,其设计宗旨是"write less,Do More",就是以最少的代码,做更多的事。它封装 JavaScript 常用的功能代码,优化 HTML 文档操作、事件处理、动画设计和 Ajax 交互,极大地简化了 JavaScript 编程,近年来深受前端开发人员欢迎。

jQuery 是一个 JavaScript 函数库,以 js 文件的形式存在,可以在官网 https://jquery.com 上下载,目前最新的版本是 3.6.0,下载其压缩版 jquery-3.6.0.min.js。在 HTML 文档中,使用<script src="\js\jquery-3.6.0.min.js"></script>代码引入 jQuery。

2.3.2 jQuery 实例

首先,如图 2.49 左侧代码所示,在<head></head>标签处引入 jquery-3.6.0.min.js,然后编写脚本。基于 jQuery 的代码比 JavaScript 的代码简洁,$(document).ready (function(){函数体})表示 DOM 加载后触发的事件,也就是当所有 HTML 标签都加载后执行函数体,$("#img1").click(function(){函数体})表示id为"#img1"的 HTML 元素的单击事件。$()表示 jQuery 对象,括号中可以是 window 对象、document 对象,大部分是通过选择器选择的 HTML 元素。$(this)表示当前的 jQuery 对象。

图 2.49　jQuery 第一个实例

2.3.3　jQuery 代码的执行顺序

jQuery 代码的执行顺序是从 `$(document).ready()` 开始的,这是 jQuery 代码执行的入口点,再通过此函数引用其他函数执行,其他函数可以是 html 元素的事件处理函数或自定义函数。

```
$(document).ready(function(){
    //jQuery 代码或者函数;
})
```

上面的代码简洁写法如下。

```
$(function(){
    //jQuery 代码或者函数;
})
```

html 元素的事件处理函数格式如下。当元素的事件触发时,调用该函数,如 `$("#img1").click()`,在 ID 值为 img1 的 html 元素被单击时触发该函数。

```
$("html 元素").事件名(function(){
    动作触发后执行的代码;
});
```

也可以写成下面的形式。

```
html 元素.事件名 = function(){
    动作触发后执行的代码;
};
```

自定义函数,可以被事件处理函数调用。

```
function 函数名(){
    函数体;
}
```

2.3.4　表单处理相关的事件、方法与实例

1. 相关的事件(表 2.2)

表 2.2　表单处理相关的事件

事件名	相关的表单元素	有关说明
submit	form	表单提交时触发
click	Button	单击普通按钮时触发
change	input type = "text" input type = "file"	文本框、文件选择框等内容改变时触发
select	select	下拉列表框被选中时触发

2. 相关的方法（表2.3）

表 2.3 表单处理相关的方法

方法名	说明
表单元素.val()	取表单元素的值，如 text1.val()
表单元素.val("")	设置表单元素的值为空，即清空表单元素值
表单元素.focus()	表单元素获得焦点
表单元素.blur()	表单元素失去焦点
alert()	弹出一个对话框

3. 相关的实例

【例2.1】 代码如图2.50所示，运行结果如图2.51所示。表单中有填写用户名的文本框、填写手机号的文本框和密码框。当点击用户名文本框时，显示相应的提示文本，表单提交时，弹出一个对话框，显示填写的各种信息，usernameDiv 输入框通过 hidden 属性设置为不可见，当单击 username 文本框时，通过 $("#usernameDiv").show() 将 usernameDiv 输入框显示出来，给出输入的提示信息，$("#username").val() 取得 username 的值，alert() 方法弹出一个对话框。

```html
<!DOCTYPE html>
<html>
    <head>
        <meta charset="utf-8" />
        <title>表单处理</title>
        <script src="js/jquery-3.6.0.min.js"></script>
        <script>
            $(document).ready(function(){
                $("#username").click(function(){
                    $("#usernameDiv").show();
                });
                $("#form1").submit(function(){
                    var username=$("#username").val();
                    var password=$("#pwd").val();
                    var phone=$("#phone").val();
                    alert("username:"+username+"\n"+"password:"+password+"\n"+"phone:"+phone);
                });
            });
        </script>
    </head>
    <body>
        <form id="form1" action="doRegister.do" method="post">
            <div id="usernameDiv" width="100px" height="50px" style="background:blue;font-size:larger;" hidden>
                用户名由英文字符和数字组成
            </div>
            用户名：<input type="text" name="username" id="username" placeholder="请输入用户名"/><br>
            手机号：<input type="text" name="phone" id="phone" placeholder="请输入手机号" /><br>
            密码：<input type="password" name="pwd" id="pwd" placeholder="请输入密码" /><br>
            <input type="submit" value="注册" />
            <input type="reset" value="重置" />
        </form>
    </body>
</html>
```

图 2.50 例 2.1 的代码

【例2.2】 代码如图2.52所示，运行结果如图2.53所示，单击表单中的 button1 按钮时清空表单数据，并弹出对话框，显示"数据已被清空"，$("#username").val("") 设置 username 的值为空。

图 2.51　例 2.1 的运行结果

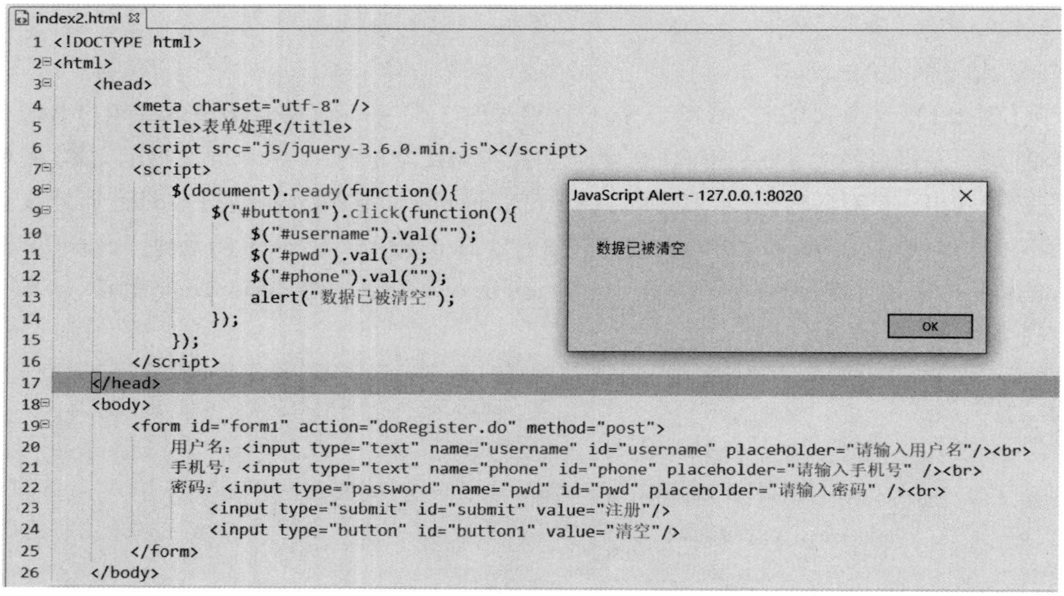

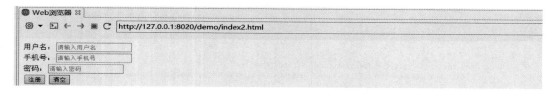

图 2.52　例 2.2 的代码

图 2.53　例 2.2 的运行结果

2.4　Bootstrap 框 架

2.4.1　Bootstrap 框架概述

2011 年 8 月，美国推特（Twitter）公司推出了开源框架 Bootstrap。Bootstrap 框架是一个基于 HTML、DIV、CSS 和 jQuery 的前端开发框架，使用 Bootstrap 可以快速地开发前端页面。

2.4.2 Bootstrap 框架的安装

使用 Bootstrap 框架，要在页面头部先进行引用，有两种方式，一种是在线的，可以使用 BootCDN 提供的免费 CDN 加速服务，引用的格式如下。

```
<script src = "http://cdn.bootcss.com/bootstrap/3.4.1/js/bootstrap.min.js"/>
<link href = "http://cdn.bootcss.com/bootstrap/3.4.1/css/bootstrap.min.css" rel = "stylesheet"/>
```

另一种是离线的，需要下载 Bootstrap 框架包，可以在 Bootstrap 中文网下载，下载的网址是 https://www.bootcss.com，得到 bootstrap-3.4.1-dist.zip 压缩包，解压后的文件夹如图2.54 所示，将其中 css 文件夹中的 bootstrap.min.css 文件拷贝到项目的 css 文件夹中，并将 js 文件夹中的 bootstrap.min.js 文件拷贝到项目的 js 文件夹中。Bootstrap 框架是基于 jQuery 实现的，所以要先将 jQuery 框架的 jquery-3.6.0.min.js 文件拷贝到项目的 js 文件夹中，引用的格式如下。

```
<script src = "js/jquery-3.6.0.min.js"/>
<script src = "js/bootstrap.min.js"/>
<link rel = "stylesheet" href = "css/bootstrap.min.css"/>
```

图 2.54 bootstrap-3.4.1-dist 目录的内容

2.4.3 使用 Bootstrap 框架进行页面的布局

Bootstrap 框架提供了一套响应式、移动设备优先的流式栅格系统。栅格系统将屏幕大小平分为 12 份，通过媒体查询得到屏幕尺寸从而可以实现响应式。采用 DIV + CSS 框架预定义类，顶级容器是 container，然后分为多个行(row)，每一行再由 12 个列(column)组成，如同表格布局。根据不同屏幕尺寸，预定义的 column 有". col – xs – * "(超小屏幕手机)、". col – sm – * "(小屏幕平板电脑)、". col – md – * "(中等屏幕桌面显示器) 和". col – lg – * "(大屏幕大桌面显示器)。* 号可以设置列，一行总的列数不能超过 12，如超过，自动掉到下一行。页面内容只能放在列 column 中，列与列之间如果需要有间隔，不紧挨着，除了设置 padding 属性(填充距离)、margin 属性(边距)，还可以使用列偏移来设置，可以设置列与列之间偏移多少列。如图 2.55、图 2.56 所示，展示了一个 Bootstrap 框架展示页的代码，运行效果如图 2.57 所示。

```
1  <!DOCTYPE html>
2  <html>
3    <head>
4      <meta charset="UTF-8">
5      <meta name="viewport" content="width=device-width,initial-scale=1" />
6      <title>bootstrap布局展示页面</title>
7      <link href="css/bootstrap.min.css" rel="stylesheet">
```

图 2.55　Bootstrap 框架展示页代码 1

```
8      <style>
9        .row{
10         margin-top:15px;
11         margin-bottom:15px;
12       }
13       [class*="col-md"]{
14         padding-left:50px;
15         padding-right:15px;
16         padding-top:15px;
17         padding-bottom:15px;
18       }
19       .col-md-12{
20         height:200px;
21         padding-top:15px;
22         padding-bottom:15px;
23         background-color:rgba(86,61,124,0.15);
24         border:1px solid rgba(86,61,124,0.2);
25       }
26     </style>
27   </head>
28   <body>
29     <div class="contrainer">
30       <div class="row">
31         <div class="col-md-12">图书管理系统</div>
32       </div>
33     </div>
34     <div class="row" style="background-color:rgba(86,61,124,0.15);padding-top:15px;padding-bottom:15px;">
35       <div class="col-md-4" >图书列表</div>
36       <div class="col-md-offset-4 col-md-4 " style="align-content:right;">   登录  |   注销</div>
37     </div>
38     <div class="row">
39       <div class="col-md-3" style="height:500px;border:1px solid rgba(86,61,124,0.2);">类别</div>
40       <div class="col-md-9" style="height:500px;border:1px solid rgba(86,61,124,0.2);">图书展示</div>
41     </div>
42   </body>
43  </html>
```

图 2.56　Bootstrap 框架展示页代码 2

图 2.57　Bootstrap 框架展示页的运行效果

2.4.4 常见网页元素的预定义类

1. jumbotron 类

超大屏幕定义了一个大的灰色背景框，可以用于显示网站名称、插入网站 logo，也可作为其他需要大块区域的容器，甚至可以将整个页面内容都放置在超大屏幕中显示，这时需要将 container 放在 jumbotron 里面。jumbotron 示例代码如图 2.58 所示，运行效果如图 2.59 所示。

```
30    <div class="container">
31        <div class="jumbotron">
32            <h1>图书管理系统</h1>
33        </div>
```

图 2.58 jumbotron 示例代码

图 2.59 jumbotron 示例代码的运行效果

2. 表格相关的预定义类

基本表格样式的类：table。
带边框的类：table table-bordered。
斑马线类：table table-striped。
鼠标悬停高亮行的类：table table-hover。
紧凑型表格的类：table table-condersed。
table 预定义类的示例代码如图 2.60 所示，运行效果如图 2.61 所示。

```
41    <table class="table table-striped table-bordered">
42        <tr>
43            <th>序号</th>
44            <th>书名</th>
45            <th>作者</th>
46            <th>价格</th>
47            <th>出版社</th>
48        </tr>
49        <tr>
56        <tr>
63        <tr>
64            <th>3</th>
65            <th>计算机组成原理（微课版）</th>
66            <th>谭志虎</th>
67            <th>55.84</th>
68            <th>人民邮电出版社</th>
69        </tr>
70    </table>
```

图 2.60 table 预定义类的示例代码

序号	书名	作者	价格	出版社
1	心理健康（第2版）	逢永花 薛敏	22.4	人民邮电出版社
2	大学生心理健康教育（慕课版 第2版）	夏翠翠	31.84	人民邮电出版社
3	计算机组成原理（微课版）	谭志虎	55.84	人民邮电出版社

图 2.61　table 预定义类示例代码的运行效果

3. 表单相关的预定义类

Bootstrap 框架提供了三种类型的表单布局，即 form（默认为垂直布局，表单标签和表单元素不在同一行）、form-inline（内联样式，所有表单元素显示在同一行）和 form-horizontal（水平布局，即表单标签和表单元素处在同一行）。

Bootstsrap 框架提供预定义类 form-control 定义表单元素的样式，表单元素的宽度为 100%。

设置边框为#ccc（浅灰色），placeholder 的颜色为#999；4 px 圆角效果；设置阴影效果，并且元素得到焦点时，阴影和边框效果有所变化。表单示例代码如图 2.62 所示，运行效果如图 2.63 所示。

```
<form class="form" role="form">
    <div class="form-group" >
        <label for="title">书名</label>
        <input type="text" class="form-control" name="title" id="title" placeholder="书名">
    </div>
    <div class="form-group" >
        <label for="type">类型</label>
        <input type="text" class="form-control" name="type" id="type" placeholder="类型">
    </div>
    <button type="submit" class="btn btn-default">查询</button>
</form>
```

图 2.62　表单示例代码

图 2.63　表单示例代码的运行效果

从图 2.62 和图 2.63 中可以看到，表单设置"form"预定义样式，即垂直布局，表单标签和表单元素不处于同一行。< div class = "form-group" ></ div >标签里面放的是同一组的表单标签和表单元素。表单元素通过 class = "form-control" 设置了预定义的样式，表单元素的宽度为 100%。如果不设置为预定义样式 form-control，则运行的效果如图 2.64 所示，表单元素宽度没有占 100%。

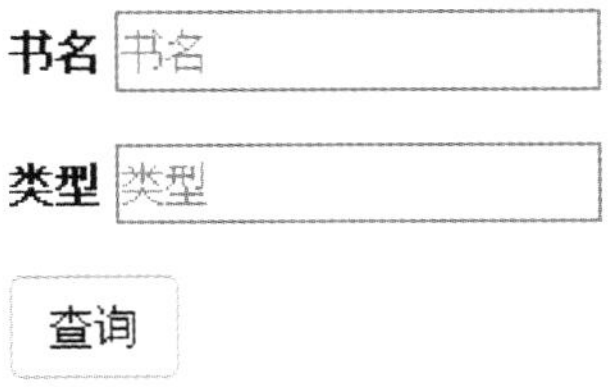

图 2.64　没有使用 form-control 的运行效果

从图 2.64 中也可以看到,在 < div class = "form-group" > </div > 标签里面的表单标签和表单元素处于同一行。

从图 2.65 和图 2.66 中可以看到,表单设置内联样式,所有的表单元素处于同一行。

图 2.65　表单示例代码

图 2.66　表单示例代码的运行效果

习　　题

1. 使用 DIV + CSS 进行网页布局,设计一个网页进行自我介绍,包括"我的学校""我的专业"和"个人简介"三个板块。

2. 使用 BootStrap 框架对图 2.16 的注册表单进行样式设置。

3. 查阅资料,使用 jQuery 的插件对图 2.16 的注册表单进行验证。

第 3 章　JSP 技术

【本章内容】

- 3.1　JSP 概述
- 3.2　JSP 基本语法
- 3.3　JSP 内置对象
- 3.4　JSP 综合案例

3.1　JSP 概述

3.1.1　什么是 JSP

　　JSP(Java Server Page)是 Sun 公司于 1999 年 6 月推出的一种基于 Java 语言的动态网页技术。它是在 Servlet 技术的基础上发展起来的,主要方式是在 HTML 文件中加入 JSP 脚本代码。JSP 脚本代码包括指令代码、Java 代码和 JSP 标签,在服务器端执行,执行结果再结合页面中的 HTML 代码,一起返回给客户端,在浏览器上显示出来。原先使用 Servlet 技术实现动态网页是通过 out.println(" <html>")的方式控制页面输出,这样显得很烦琐,而使用 JSP 技术能够很方便地控制页面的输出。

3.1.2　JSP 程序的运行机制

　　用户访问 JSP 文件,服务器找到相应的文件,将 JSP 转译成 Servlet,再将 Servlet 编译为字节码,然后将字节码加载到内存并运行,运行后生成的 HTML 代码再发送给客户端,在浏览器上显示页面。JSP 程序的运行机制如图 3.1 所示。

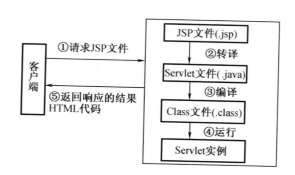

图 3.1　JSP 程序的运行机制

【例3.1】第一个JSP文件。

```jsp
<%@ page language="java" contentType="text/HTML;charset=utf-8"%>
<!DOCTYPE html>
<html>
<head>
<title>3.1</title>
</head>
<body>
<%
    out.println("第一个JSP文件!");
%>
</body>
</html>
```

3.2 JSP基本语法

3.2.1 JSP文件基本组成

一个完整的JSP文件是由HTML元素、JSP指令元素、JSP脚本元素、JSP动作元素和注释元素所组成的,如例3.2所示。

【例3.2】 JSP文件组成示例。

```jsp
<%@ page language="java" import="java.util.*" contentType="text/HTML;charset=utf-8"%>
<%@ taglib uri="http://java.sun.com/jsp/jstl/core" prefix="c"%>
<!DOCTYPE html>
<html>
<head>
<title>3.2</title>
</head>
<body>
<%
    //在页面输出"hello,jsp"
    out.println("hello,jsp!");
%>
服务器端的时间是:<%=new Date()%>
<hr>
<c:out value="全局变量的演示"/>
    //声明元素,声明sum是一个全局变量
    <% int sum=0;%>
```

```
    %>
    <%
      for(int i=0;i<5;i++){
          sum=sum+1;
      }
      out.println("sum="+sum+"<br>");
    %>
    </body>
</html>
```

3.2.2 HTML 元素

HTML(HyperText Markup Language,超文本标记语言)是网页制作的基础语言。HTML 元素由各种标记组成,形式如<html></html>,用来标识各种网页元素,如文本、图片、音频、视频等,由浏览器解释执行,网页及各种资源文件通过超链接构成万维网。

3.2.3 JSP 指令元素

JSP 指令元素用于指示 JSP 容器如何处理 JSP 页面,JSP 容器会根据指令信息进行编译,生成 Java 文件。

1. page 指令

page 指令用于设置 JSP 页面的属性,通常放在一个 JSP 文件的第一行位置,其格式如下。

```
<%@page 属性名="属性值" 属性名="属性值" %>
```

page 指令总共有十多个属性,下面列出常用的属性:

①language:定义页面中服务器端代码所采用的编程语言,默认是 Java。

②pageEncoding:指定页面的字符编码,默认是 ISO-8859-1,如果需要支持中文字符,则要设置为 GBK 或 GB2312 或 UTF-8,也可以在 IDE 中进行统一设置。pageEncoding="GB2312" 或 pageEncoding="GBK" 或 pageEncoding="UTF-8"。

③contentType:设置客户端响应的类型和字符编码,用于告诉浏览器服务器返回内容的类型及编码格式,类型采用 MIME(Multipurpose Internet Mail Extensions,多用途互联网邮件扩展类型),这种类型最早应用于电子邮件系统,后来也应用于浏览器。服务器发送的多媒体数据类型可以是 Word 文档、Excel 文档、图像、音频、视频等,浏览器会使用相应的插件来打开文件,如果没有相应的插件,则提示用户将接收到的内容另存为文件。

通常响应的是网页,所以设置为:

```
contentType="text/HTML;charset=GB2312"
```

下面的代码表示响应的是 xlsx 类型的文档。

```
contentType="application/vnd.openxmlformats-officedocument.spreadsheetml.sheet"
```

④import：导入页面要使用到的类和包，各类或包之间用逗号隔开或分多行书写。

```
<%@page import="java.util.*"%>
<%@page import="java.util.*,java.sql.*"%>
```

⑤errorPage 和 isErrorPage：成对一起使用，设置一个页面捕获异常，并对异常做出处理。

```
<%@page errorPage="用于捕获异常的页面"%>
<%@page isErrorPage="true|false"%>
```

2. taglib 指令

该指令指定在 JSP 页面中使用的扩展标签，可以是 JSTL（标准标签库）中的标签，也可以是用户自定义的标签，通常情况下是使用 JSTL 定义的标签。JSTL 包含 5 个库，分别是核心标签库、格式化标签库、函数标签库、SQL 标签库和 XML 标签库，具体在后面 JSTL 部分介绍。该指令格式如下：

```
<%@taglib uri="http://java.sun.com/jsp/jstl/core" prefix="c"%>
```

uri 指明标签库的位置，prefix 指明在 JSP 页面中使用的前缀。指定使用哪个标签库中的标签，那么在 JSP 文件中就可以利用标签实现一定的功能，如：

```
<c:out value="12"/>
```

该代码使用 out 标签在页面中输出 12。

3. include 指令

该指令用来包含文件，这种包含属于静态包含，是在转译时将被包含文件的内容包含进来后再执行，所以文本文档里面的 JSP 代码被包含进来后被执行。

格式：`<%@include file="被包含文件的路径和文件名"%>`

3.2.4　JSP 脚本元素

JSP 脚本元素包括声明元素、表达式和 Java 代码段。

1. 声明元素

格式：`<%! int i=0;%>`

声明的元素作为全局变量。

【例 3.3】　声明元素的运用。

```
<%@page language="java" pageEncoding="UTF-8"%>
<!DOCTYPE html>
<html>
<head>
<meta http-equiv="Content-Type" content="text/html; charset=UTF-8">
<title>Insert title here</title>
</head>
<body>
<%! int count=0;%>
<%
count=count+1;
out.print("当前的点击数是："+count);
```

```
%>
</body>
</html>
```

通过<%!int count=0;%>声明 count 是一个全局变量,在整个应用程序运行期间都有效,可以用于统计系统运行时的点击数,打开浏览器运行一次,count 加 1,当服务器关闭或重启时,count=0。

2. 表达式

格式:`<%=1+2%>`

注意:后面没有分号,表达式的作用是先计算,再输出。

3. Java 代码段

Java 代码段又称脚本代码,是使用<% %>括起来的一段 Java 程序,遵循 Java 的语法,在 JSP 文件中被大量使用。

3.2.5 JSP 动作元素

JSP 动作元素包括 forward 元素、include 元素和 plugin 元素,形式如 HTML 标签。

1. forward 元素

forward 元素实现转发,从服务器端转发到另一个页面或程序。

格式:`<jsp:forward page="被包含的文件名"></jsp:forward>`

转发时可以传递参数,通过`<jsp:param name="参数名" value="参数值"></jsp:param>`来实现。

【例 3.4】 forward 元素的应用示例。

forward.jsp 文件:

```
<%@page language="java" pageEncoding="UTF-8"%>
<!DOCTYPE html>
<html>
<head>
<meta http-equiv="Content-Type" content="text/html; charset=UTF-8">
<title>forward.jsp</title>
</head>
<body>
```

下面代码实现转发,转到页面 a.jsp,同时可以通过 param 元素设置转发时传递的参数。

```
<jsp:forward page="a.jsp">
    <jsp:param name="age" value="12"></jsp:param>
</jsp:forward>
</body>
</html>
```

a.jsp 文件:

```
<%@page language="java" pageEncoding="UTF-8"%>
<!DOCTYPE html>
```

```
<html>
<head>
<meta http-equiv = "Content-Type" content = "text/html; charset =UTF-8" >
<title>a.jsp</title>
</head>
<body>
a 页面的内容,传递过来的参数 age = <% = request.getParameter("age") % >
</body>
</html>
```

上面的代码通过 request.getParameter("age")取得参数的值。从图 3.2 中可以看到,地址栏显示的文件名是 forward.jsp,页面上显示的是 a.jsp 文件的内容,说明是由 forward.jsp 在服务器端转发,转到 a.jsp 文件,同时通过 <jsp:param> 传递 age 参数。

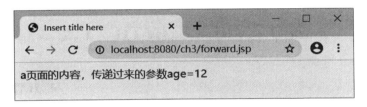

图 3.2　forward 实现转发

2. include 元素

include 元素包含文件,这种包含是动态包含,是在执行的时候包含进来的。如果被包含文件是文本文档,里面有 JSP 代码,则这些 JSP 代码是不会被执行的,所以看不到所要的结果。

格式:<jsp:include page = "被包含的文件名" > </jsp:include >

3. plugin 元素

plugin 元素用得比较少,这里不做介绍。

3.3　JSP 内置对象

3.3.1　内置对象概述

任何一种动态网页技术都有内置对象,Web 服务器依靠内置对象实现服务器的功能。所谓内置对象就是这些对象在服务器启动时就会被创建,可以直接使用,而不需要通过 new 来创建。JSP 内置对象主要有 9 种,这些内置对象所依赖的类或接口如表 3.1 所示。

表 3.1　JSP 内置对象及其所依赖的类或接口

内置对象	依赖的类或接口
out	javax. servlet. jsp. JspWriter
request	javax. servlet. http. HttpServletRequest
response	javax. servlet. http. HttpServletResponse
session	javax. servlet. http. HttpSession
application	javax. servlet. ServletContext
exception	java. lang. Trowable
config	javax. servlet. ServletConfig
page	java. lang. Object
pageContext	javax. servlet. jsp. PageContext

1. request 对象、response 对象和 session 对象与服务器的交互

request、response 和 session 是 3 个重要的 JSP 内置对象,下面从图 3.3 客户端与服务器交互认识请求、响应和会话的过程。

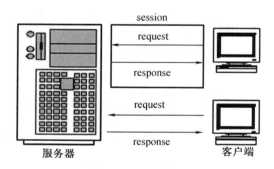

图 3.3　客户端与服务器的交互

当用户打开客户端浏览器,在地址栏中输入服务器 Web 页面的地址后,客户端浏览器就会显示相应的页面。从图 3.3 中可以看到,客户端和服务器进行的是请求和响应的交互过程,客户端浏览器从 Web 服务器上获得网页,实际过程是使用 HTTP 协议向服务器发送了一个请求,服务器在收到来自客户端浏览器发来的请求后响应请求。JSP 通过 request 对象获取客户端浏览器的请求,通过 response 对象对客户端浏览器进行响应,而 session 对象则一直保存着会话期间所需要传递的数据信息。

2. HTTP 协议

HTTP 协议,即超文本传输协议,是建立在 TCP/IP 之上的协议。该协议规定了客户端浏览器如何向 Web 服务器请求资源,服务器如何对客户端的请求做出响应,是一种无状态协议。当客户端浏览器要从服务器获取页面时,浏览器首先建立一个对服务器的连接,并发出请求,服务器收到请求后响应,所以 HTTP 协议又被称作"请求和响应"协议。

(1) HTTP 请求

HTTP 请求具有某种结构,具体包括请求行、请求头和可能的信息体,具体如下:

```
GET/hello.htm HTTP/1.1              请求行
Host:www.sina.com.cn                请求头
⋮
信息体
```

①首行是请求行,规定了请求的方法、请求的资源及使用的 HTTP 协议的版本。

②第二行是请求头,一个请求可以包含多个请求头,提供了关于信息体的附加信息及请求的来源。其中,Host 头规定了请求页面的 Internet 地址。

③一个请求还可能包含信息体,如包含通过表单提交的内容。

(2) HTTP 响应

HTTP 响应也具有某种结构,包括状态行、响应头及可能的信息体,这些响应信息由服务器发送给客户端的浏览器。

```
HTTP/1.1 200 OK                              状态行
Content - Type:text/html;charset = ISO - 8859 - 1    响应头
⋮
信息体
```

①状态行说明了正在使用的协议、状态代码及文本信息。

②一个响应可以包含多个响应头,如 Refresh 响应头就是告诉浏览器要定时刷新。

③信息体是客户端请求的网页的运行结果,JSP 页面是执行后的网页的静态信息。

3.3.2 out 对象

out 对象用于实现服务器向客户端浏览器输出信息,也用于实现对服务器的输出缓冲区进行管理。out 对象的基类是 javax.servlet.jsp.JspWriter 类。JspWriter 对象输出数据时,不是直接输出到浏览器,而是先将数据保存在服务器的输出缓冲区中,然后再根据要求进行输出。out 对象常用的方法有 println( )、print( )、flush( )、close( )、clear( )等,如表 3.2 所示。

表 3.2 out 对象常用的方法

方法名	实现的功能
println( )或 print( )	向客户端浏览器输出数据,println( )会在输出内容后面添加换行标记
flush( )	将缓冲区的数据输出到客户端浏览器
close( )	关闭输出流,后面的代码不会被执行
clear( )	清除缓冲区数据,但没有将数据输出到客户端浏览器
clearBuffer( )	清除缓冲区数据,清除之前先把数据输出到客户端浏览器
getBufferSize( )	获得缓冲区的大小
getRemaining( )	获得缓冲区剩余空间的大小
isAutoFlush( )	判断缓冲区是否自动输出

对服务器的输出缓冲区进行管理要结合 page 指令中的 buffer 和 autoFlush 两个属性。

```
<%page buffer = "none |default |sizekb"% >
```

设置 out 对象服务器缓冲区的大小,buffer = "none",设置无缓冲区,则 out 对象将数据直接输出到客户端浏览器;buffer = "default",设置缓冲区大小是默认值 8 kb,也可以根据实际需要设置缓冲区的大小。

```
<%page autoFlush = "true |false" >
```

设置缓冲区是否自动刷新,默认值为"true",则缓冲区满时自动刷新。当 autoFlush 设置为"false",需要刷新时,则使用 out 对象的 flush 方法实现刷新。

3.3.3 request 对象

request 对象用于封装客户端向服务器提交的数据,包括客户端提交的数据和请求时附加的客户端信息及请求资源的信息,通过它提供的方法和属性可以获取这些信息。另外,也可以通过设置属性的方式在一次请求期间进行数据的传递。

1. 通过 request 对象获取表单元素的值

【例 3.5】 request 对象获取表单元素的值示例。

form. jsp 文件:

```
< form name = "form1" method = "post" action = "do_form1.jsp" >
用户名:< input type = "text" name = "username"/ > < br >
密码:< input type = "password" name = "password"/ > < br >
个人简介:< textarea name = "introduce" rows = "3" cols = "30" >
        < /textarea > < br >
性别:< input type = "radio" name = "sex" value = "男" >男
< input type = "radio" name = "sex" value = "女" >女< br >
爱好:< input type = "checkbox" name = "hobby" value = "运动" >运动
      < input type = "checkbox" name = "hobby" value = "购物" >购物
      < input type = "checkbox" name = "hobby" value = "美食" >美食
      < input type = "checkbox" name = "hobby" value = "读书" >读书< br >
所在院系:< select name = "department" >
        < option value = "数学系" >数学系< /option >
        < option value = "计算机系" >计算机系< /option >
        < option value = "艺术系" >艺术系< /option >
    < /select > < br >
  < input type = "submit" value = "提交"/ >
  < input type = "reset" value = "重置"/ >
< /form >
```

do_form1. jsp 文件,通过 getParameter()获得指定参数的值,通过 getParameterValues()来获得指定参数的多个值,然后通过 for 循环取得每一个值。

do_form1. jsp 文件:

```jsp
<body >
<%
    request.setCharacterEncoding("utf-8");
    String username = request.getParameter("username");
    String sex = request.getParameter("sex");
    String department = request.getParameter("department");
    String introduce = request.getParameter("introduce");
    String[] hobbies = request.getParameterValues("hobby");
    out.print("姓名:" + username + "<br >");
    out.print("性别:" + sex + "<br >");
    out.print("所在院系:" + department + "<br >");
    out.print("有" + hobbies.length + "项爱好:" + "<hr >");
    for(int i = 0;i < hobbies.length;i + +){
        out.print(hobbies[i]);
    }
%>
</body >
```

form.jsp 和 do_form1.jsp 运行结果如图 3.4 所示。

图 3.4　**form.jsp 和 do_form1.jsp 运行结果**

2. 通过 request 对象获取客户端的信息和请求资源的信息

通过 request 对象可以获得请求时附加的客户端信息和请求资源的信息,如使用的访问协议、客户端 IP 地址、客户端计算机名称、请求资源的 URL,也可以得到请求报头的各个参数的值。具体方法如下:

getSchema():返回当前请求页面使用的协议。

getServerName():返回请求页面所在的服务器的名称。

getServerPort():返回请求页面所在的服务器的端口号,默认是 80。

getContextPath():返回请求的应用系统的名称。

getRemoteAddr():返回应用系统的客户端的 IP 地址。

getRemoteHost():返回客户端的计算机名称。

getServletPath()：返回 Servlet 程序的路径。

getRealPath()：返回应用系统的实际路径。

getQueryString()：返回 URL 地址中的查询字符串。

getRequestURL()：返回请求资源的 URL。

getHeader()：返回请求报头的参数值，并指明参数名。

getHeaderNames()：返回所有报头的名称。

getRequestDispatcher()：获得 request 转发器，利用转发器的 forward()方法可以实现转发。

【例 3.6】 getHeaderNames()的例子。

```
<body>
下面显示请求报头的名称和信息：
<hr>
<%
    Enumeration names = request.getHeaderNames();    //读取所有请求报头的名称,并
                                                      //保存在枚举型对象中

    while(names.hasMoreElements()){
        String name = (String)names.nextElement();    //取出每一个请求报头的名称
        String value = request.getHeader(name);       //读取指定请求报头的信息
        out.println(name + ":  " + value + "<br>");   //输出请求报头信息
    }
%>
</body>
```

运行结果如图 3.5 所示。

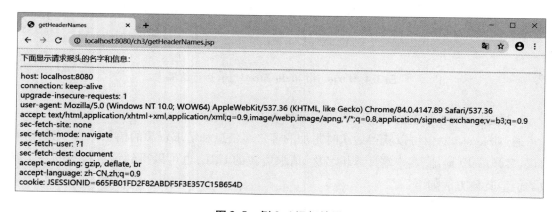

图 3.5 例 3.6 运行结果

【例 3.7】 getContextPath()的例子。

```
String  path = request.getContextPath();    //得到请求的应用系统名称
String  basePath = request.getScheme() + "://" + request.getServerName() + ":" + request.getServerPort() + "/" + path + "/";
<base href = " <% = basePath% > " >   //应用系统的基础路径
```

3. 通过设置属性的方式保存 request 作用域的数据

setAttribute(String name,Object value):设置 request 作用域的属性。

getAttribute(String name):返回 request 作用域指定属性的值,返回值是 Object 类型。

getAttributeNames():返回 request 作用域所有属性的值,返回值是一个枚举型。

removeAttribute(String name):移除 request 作用域指定属性。

【例 3.8】 使用 request 设置属性的方式来传递数据。

requestAttribute1.jsp 文件:

```
<body>
使用 request 设置属性的方式来传递数据
<%
request.setAttribute("name","Jack");
request.getRequestDispatcher("requestAttribute2.jsp").forward(request,response);  //服务器端跳转
//response.sendRedirect("requestAttribute2.jsp");
%>
</body>
```

requestAttribute2.jsp 文件:

```
<body>
验证是否获得 request 属性的值
<%
String name = (String)request.getAttribute("name");
out.println("name = " + name);
%>
</body>
```

运行结果如图 3.6 所示,左边的结果是在 rquestAttribute1.jsp 文件中,设置 request 作用域属性后,再通过 request.getRequestDispatcher("requestAttribute2.jsp")获取转发器,然后实现转发,直接在服务器端转向执行 requestAttribute2.jsp 文件,只有一次请求,name 属性的值在一次请求范围内有效,所以得到 name = Jack。当我们把语句改为 response.sendRedirect()方法,执行的是重定向,是由客户端重新发出一次请求,地址栏可以看到显示的是"requestAttribute2.jsp"地址,是两次请求,在上一次请求的 name 的值没办法传递到下一次请求,所以页面上显示 name = null。

图 3.6　例 3.8 运行结果

3.3.4　response 对象

response 对象用于实现由服务器端向客户端发送信息,常用的方法如下:

getWriter():返回一个 PrintWriter 对象,由它输出字符信息,实现的功能跟 out 对象一样。

getOutputStream():返回一个 ServletOutputStream 输出流对象,由它输出二进制信息。

setContentType():设置响应内容的 MIME 类型及字符编码。

setCharacterEncoding():设置字符编码。

setHeader(String s,String value):给指定的响应头设置值。

setHeader("refresh","2;url=1.jsp"):设置刷新,2 s 后自动跳转到 1.jsp 页面。

sendRedirect("2.jsp"):重定向到 2.jsp 页面,由服务器将重定向的地址发送给客户端,客户端浏览器重新发送请求。

3.3.5　cookie

cookie 技术是服务器将数据保存到客户端浏览器内存或硬盘文件中的一种技术。通过 cookie,服务器可以将用户访问网站时的一些信息记录下来,可实现用户快捷登录或为用户提供一些个性化的服务。cookie 保存数据是以键值对即 name=value 的方式保存的,保存的数据是有有效时间的,根据设置的值暂存在浏览器内存中或以文件的形式保存在硬盘的某个位置。

1. 创建 cookie 对象

通过 javax.servlet.http.Cookie 类来创建一个 cookie 对象,格式如下:

```
Cookie cookie1 = new Cookie(String name,String value)
```

创建 cookie 对象时要指明键值对,第一个参数 name 是键名,第二个参数 value 是键值。

2. 将数据写入 cookie

通过 response 对象的 addCookie()方法添加 cookie,将数据保存到客户端。

addCookie(Cookie cookie):添加 cookie。

添加 cookie 之前先要设置 cookie 的生命周期。

setMaxAge(int expiry):设置生命周期。

有效时间以秒为单位,默认值为 -1,负数表示这个 cookie 对象是临时的,暂存在浏览器内存中,关闭浏览器数据就会丢失。如果生命周期为 0,表示删除这个 cookie 对象。

3. 将数据从 cookie 中读取出来

读取 cookie 是通过 request 对象的 getCookies() 完成的。

getCookies():将所有 cookie 对象读取出来。

可以通过两个方法取得键名和键值。

getName():返回 cookie 的键名。

getValue():返回 cookie 的键值。

【例 3.9】 定义两个 cookie 对象,用于保存 name 的值和 age 的值,设置生命周期为 7 天。首先创建 writeCookie.jsp 文件,主要代码如下:

```
<%
  Cookie cookie1 = new Cookie("name","jack");
  cookie1.setMaxAge(60 * 60 * 24 * 7);      //设置生命周期为7天,分开写,比较清楚
  response.addCookie(cookie1);
  Cookie cookie2 = new Cookie("age","12");
  cookie1.setMaxAge(60 * 60 * 24 * 7);
  response.addCookie(cookie2);
  out.print("写入两个cookie,生命周期为7天")
%>
```

接下来创建 getCookies.jsp,主要代码如下:

```
<%
  Cookie cookies[] = request.getCookies();
  for(int i = 0;i < cookies.length;i + +){
      String cookieName = cookies[i].getName();
      String cookieValue = cookies[i].getValue();
      out.println(cookieName + " = " + cookieValue);
  }
%>
```

先要运行 writeCookie.jsp,将数据写入 cookie,然后通过 getCookies.jsp 将数据读取出来,输出在页面,运行结果如图 3.7 所示。

图 3.7 例 3.9 的运行结果

可以看到除了显示设置的两个键值,最后还显示了当前的 sessionID,标识当前用户。

3.3.6 session 对象

session 对象用来保存一个用户一次会话期间的数据。一次会话指的是一个用户从他

打开浏览器访问 Web 应用程序的一个页面开始,到关闭浏览器这段时间。在这期间,用户打开不同的页面,在不同页面之间的数据共享是通过 session 对象来实现的。为了跟踪用户,服务器给每个访问该应用程序的用户分配了一个 ID 号,通过这个 ID,可以标识不同用户。

1. 通过设置属性的方式保存 session 作用域的数据

setAttribute(String name, Object value):设置 seesion 作用域的属性。

getAttribute(String name):返回 seesion 作用域指定属性的值,返回值是 Object 类型。

getAttributeNames():返回 session 作用域所有属性的值,返回值是一个枚举型。

removeAttribute(String name):移除 session 作用域指定属性。

【例 3.10】使用 session 设置属性的方式来传递数据。

sessionAttribute1.jsp 文件:

```
<body>
  使用session设置属性的方式来传递数据
  <%
    session.setAttribute("name","Jack");
  %>
</body>
```

sessionAttribute2.jsp 文件:

```
<body>
  验证是否获得session属性的值
  <%
    String name =(String)session.getAttribute("name");
    out.println("name = " +name);
  %>
</body>
```

在 sessionAttribute1.jsp 文件中,通过设置 session 作用域属性的方式将 name 的值 Jack 保存下来,在一次会话期间有效,也就是在用户打开浏览器跳到其他页面时都有效,所以在 sessionAttribute2.jsp 文件中,可以通过 session.getAttribute("name")的方式获取到 name 的值。通过这种方式,在登录时,可以将登录的用户名保存下来,在其他页面上显示。例 3.10 运行结果如图 3.8 所示。

(a)　　　　　　　　　　　　(b)

图 3.8　例 3.10 运行结果

2. 与 session 对象有关的几个方法

isNew( ):判断 session 对象是否是新创建的,返回 true 时,表示这个 session 对象是新创建的。

getId( ):返回当前 session 对象的 ID,以字符串的形式显示。在一次会话期间,返回的 ID 是相同的。

getCreationTime( ):返回 session 对象创建的时间,是一个 LONG 型的整数,表示从格林尼治时间 1970-1-1 00:00:00 到当前时间所经历的毫秒数,可以通过转换获得具体的日期和时间。

getLastAccessTime( ):返回客户端最后一次请求时间,是一个 LONG 型的整数,表示从格林尼治时间 1970-1-1 00:00:00 到当前所经历的毫秒数。

setMaxInactiveInterval(int interval):设置 session 对象的超时时间,单位是 s。如果用户发出请求,超过 interval 设置的时间,不再向服务器发出请求,则认为 session 超时,将删除本次的 session 对象。服务器默认设置的超时时间是 30 min,如果设置为负数,表示永不超时。

invalidate( ):使 session 对象失效,使用这个方法可实现注销功能。

3. 使 session 对象失效的 4 种方式

(1) 关闭客户端浏览器窗口;

(2) 通过调用 session.invalidate( ) 使其失效;

(3) 如果用户超过设置的超时时间没发出请求,即删除 session 对象;

(4) 服务器关闭或重启。

【例 3.11】 session 对象常用属性的示例。

sessionInfor1.jsp 文件:

```
<html>
 <head>
  <title>sessionInfor1.jsp</title>
 </head>
<body>
得到 session 对象的有关信息【页面 1】:
  <hr>
  <%
  out.println("isNew(): " +session.isNew() +"<br>");
  out.println("getId(): " +session.getId() +"<br>");
  out.println("getCreationTime():" +
    session.getCreationTime() +"<br>");
  out.println("getMaxInactiveInterval():" +
    session.getMaxInactiveInterval() +"<br>");
  %>
 </body>
</html>
```

sessionInfor2.jsp 文件:

```
<html>
 <head>
  <title>sessionInfo2.jsp</title>
```

```
</head>
<body>
得到session对象的有关信息【页面2】：
<hr>
<%
   out.println("isNew():  "+session.isNew()+"<br>");
   out.println("getId():  "+session.getId()+"<br>");
   out.println("getCreationTime():" +
     session.getCreationTime() +"<br>");
   out.println("getMaxInactiveInterval():" +
     session.getMaxInactiveInterval() +"<br>");
%>
   <hr>
接下来让session失效
<%
   session.invalidate();
%>
</body>
</html>
```

sessionInfo3.jsp 文件：

```
<html>
 <head>
   <title>sessionInfo3.jsp</title>
 </head>
 <body>
得到session对象的有关信息【页面3】：
   <hr>
   <%
    out.println("isNew():  "+session.isNew()+"<br>");
    out.println("getId():  "+session.getId()+"<br>");
    out.println("getCreationTime():" +
      session.getCreationTime() +"<br>");
    out.println("getMaxInactiveInterval():" +
      session.getMaxInactiveInterval() +"<br>");
   %>
 </body>
</html>
```

在一个浏览器窗口按照顺序运行这三个文件,得到的结果如图3.9所示。如果期间关闭浏览器窗口或运行顺序不对,那么得到的结果是不相同的。

(a)　　　　　　　　　　　(b)　　　　　　　　　　　(c)

图 3.9　三个 session 文件的运行结果

从图 3.9 中可以看到,第一个页面和第二个页面的 sessionID 是相同的,说明从打开浏览器开始,都是同一次会话,第一个页面是新创建的 session,在第二个页面中设置了时间让 session 失效,所以在第三个页面中可以看到 session 是新创建的,是另一次会话,它的 sessionID 和前面两个页面是不相同的。

3.3.7　application 对象

application 对象是应用程序级的对象,用来保存应用程序级的数据,也可以通过 application 对象读取系统信息。

1. 通过设置属性的方式保存应用程序级的数据

setAttribute(String name,Object value):设置 applicaion 对象的属性,有效范围是整个应用程序。

getAttribute(String name):得到指定属性的值,返回值是一个 Object 类型。

removeAttribute(String name):移除 application 对象指定的属性。

2. 读取 servlet 容器信息

getServerInfo():返回服务器的名称和版本。

getMajorVersion():返回 Servlet api 的主版本号。

getMinorVersion():返回 Servlet api 的次版本号。

getRealPath(String path):取得指定路径的实际路径。

【例 3.12】　通过 application 对象得到系统的有关信息。

applicationInfo.jsp 文件:

```
通过application对象得到系统的有关信息:
<hr>
<%
  out.println("返回当前servlet容器的名字与版本号getServerInfo():" +
    application.getServerInfo() + "<br>");
  out.println("返回servlet api的主版本号getMajorVersion():" +
    application.getMajorVersion() + "<br>");
  out.println("返回servlet api的次版本号getMinorVersion():" +
    application.getMinorVersion() + "<br>");
  out.println("返回MIME的类型getMimeType():" +
    application.getMimeType("1.docx") + "<br>");
```

```
out.println("返回请求的应用系统的名称 getContextPath():" +
    application.getContextPath() + "<br>");
out.println("返回请求的应用系统的实际路径 getRealPath():" +
    application.getRealPath("/"));
%>
```

运行的结果如图 3.10 所示。

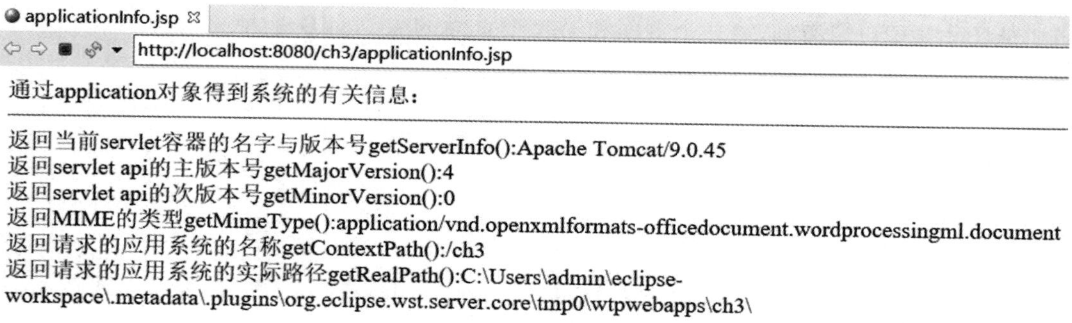

```
● applicationInfo.jsp ⊠
⇦ ⇨ ■ ⊙ │ http://localhost:8080/ch3/applicationInfo.jsp
通过application对象得到系统的有关信息：

返回当前servlet容器的名字与版本号getServerInfo():Apache Tomcat/9.0.45
返回servlet api的主版本号getMajorVersion():4
返回servlet api的次版本号getMinorVersion():0
返回MIME的类型getMimeType():application/vnd.openxmlformats-officedocument.wordprocessingml.document
返回请求的应用系统的名称getContextPath():/ch3
返回请求的应用系统的实际路径getRealPath():C:\Users\admin\eclipse-
workspace\.metadata\.plugins\org.eclipse.wst.server.core\tmp0\wtpwebapps\ch3\
```

图 3.10 例 3.12 的运行结果

3. 读取 Web 应用的初始化参数

在整个应用系统的配置文件 web.xml 中，可以设置一些全局的初始化参数，如数据库的连接参数，然后通过 getInitParameter(String name) 可以得到指定初始化参数的值，通过 getInitParameterNames() 返回所有初始化参数的名称，并保存在枚举类型对象中。

【例 3.13】 取得应用系统的所有初始化参数的名称和值。首先，在 web.xml 文件中设置连接数据库的驱动器名称、用户名和密码，然后通过 getInitParameterNames() 取得这些参数的名称和值。

web.xml 文件部分内容：

```
<context-param>
    <param-name>DRIVER</param-name>
    <param-value>com.mysql.cj.jdbc.Driver</param-value>
</context-param>
<context-param>
    <param-name>USER</param-name>
    <param-value>root</param-value>
</context-param>
<context-param>
    <param-name>PASSWORD</param-name>
    <param-value>123456</param-value>
</context-param>
```

getInitParameterNames.jsp 文件：

```jsp
<%
out.println("以下是应用系统的初始化参数:");
out.println("<hr>");
Enumeration e = application.getInitParameterNames();
while(e.hasMoreElements()){
    String name =(String)e.nextElement();
    String value = application.getInitParameter(name);
    out.println(name + ": " +value + "<br>");
}
%>
```

运行结果如图 3.11 所示。

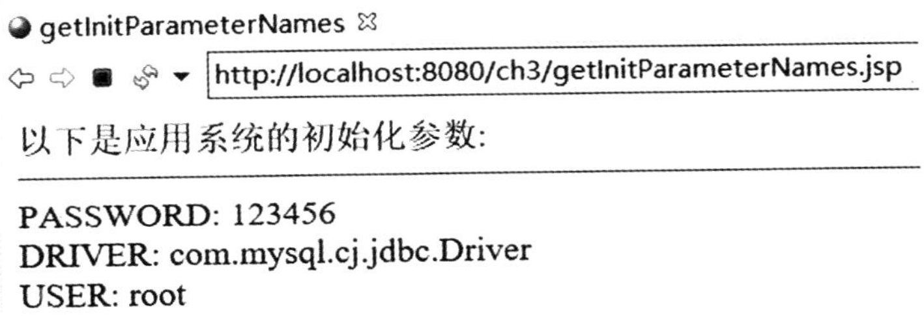

图 3.11　例 3.13 的运行结果

3.4　JSP 综合案例

本案例使用 session 对象和 application 对象统计在线人数和显示在线人员名单。用户通过登录页面输入账号和密码后进入主页面，在主页面上显示在线人数和在线人员名单。

首先，创建项目 online，并且在 webapp 目录中创建 4 个 JSP 文件，分别是 login.jsp、doLogin.jsp、online.jsp 和 invalidate.jsp，如图 3.12 所示。

1. login.jsp

```jsp
<body>
<form name="loginForm" method="post" action="doLogin.jsp">
  username:<input type="text" name="username"><br>
  <input type="submit" value="login">
</form>
</body>
```

在 login.jsp 页面输入用户名，然后点击登录按钮，表单提交给 doLogin.jsp 进行处理。

```
online
  JAX-WS Web Services
  JRE System Library [JavaSE-12]
  src/main/java
  Apache Tomcat v9.0 [Apache Tomcat v9.0]
  Deployment Descriptor: online
  build
  src
    main
      java
      webapp
        META-INF
        WEB-INF
        doLogin.jsp
        invalidate.jsp
        login.jsp
        online.jsp
```

图 3.12 online 项目目录

2. doLogin.jsp

```jsp
<body>
<%
String username = request.getParameter("username");
if(!username.equals("")){
  session.setAttribute("loginName",username);
  if(application.getAttribute("onlineCount") == null){
    //第一个访问用户
    application.setAttribute("onlineCount",1);
    //保存为哈希表的形式,记录用户的sessionId,方便后面的注销
    Map<String,String> loginUsers = new HashMap<String,String>();
    loginUsers.put(session.getId(),username);
    application.setAttribute("loginUsers",loginUsers);
  }else{
    //不是第一个访问用户
    int onlineCount = (Integer)application.getAttribute("onlineCount");
    application.setAttribute("onlineCount",onlineCount+1);
    Map<String,String> loginUsers = (HashMap<String,String>)application.
    getAttribute("loginUsers");
    loginUsers.put(session.getId(),username);
    application.setAttribute("loginUsers",loginUsers);
  }
  response.sendRedirect("online.jsp");
}
```

```
%>
</body>
```

3. online.jsp

```jsp
<body>
<%
 if(session.getAttribute("loginName")!=null){
   out.println(session.getAttribute("loginName")+",您好!"+"<br>");
  }
%>
当前在线人数:<%=application.getAttribute("onlineCount")%><br>
在线人员名单
<hr>
<%
 Map<String,String> loginUsers = (HashMap<String,String>)application.getAttribute("loginUsers");
 //哈希表要循环输出,需要转换为集合SET,然后使用for-each循环增强的方式逐个输出
 Set<Map.Entry<String,String>> entrys = loginUsers.entrySet();
 for(Map.Entry entry:entrys){
   //out.println(entry.getKey()+"  "+entry.getValue()+"<br>");
   out.println(entry.getValue()+"<br>");
  }
%>
<hr>
<a href="invalidate.jsp">注销</a>
</body>
```

online.jsp 页面显示当前登录的用户名、在线人数和在线人员名单。

4. invalidate.jsp

```jsp
<body>
<%
 String sessionId = session.getId();
 Map<String,String> loginUsers = (HashMap<String,String>)application.getAttribute("loginUsers");
 loginUsers.remove(session.getId());
 session.invalidate();
 response.sendRedirect("doLogin.jsp");
%>
</body>
```

完成代码编写后应进行测试,打开三种浏览器,如360浏览器、IE浏览器和谷歌浏览

器,模仿三个用户打开登录页面,如图3.13所示。

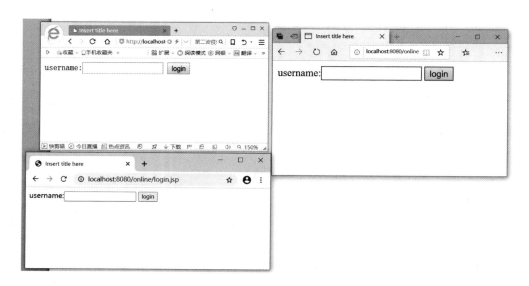

图 3.13 登录界面

在谷歌浏览器中输入用户名 a,在图 3.14 中显示"当前在线人数:1"和"在线人员名单 a"的信息。

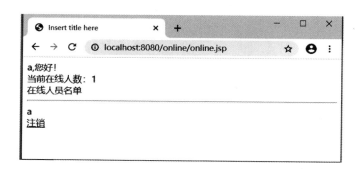

图 3.14 用户 a 已登录

在 IE 浏览器中输入用户名 b,则显示"当前在线人数:2"和"当前在线人员名单 b a",如图 3.15 所示,刷新谷歌浏览器,可以看到两个浏览器都显示出相同的信息,即"当前在线人数:2"和"在线人员名单 b a",如图 3.16 所示。

接着,在 360 浏览器中输入用户名 c,完成登录,则显示"当前在线人数:3"和"当前在线人员名单 c b a",如图 3.17 所示。

图 3.15　IE 浏览器显示的信息

图 3.16　谷歌浏览器显示的信息

图 3.17　用户 c 登录

刷新三个浏览器,如图 3.18 所示,可以看到三个浏览器都能正确显示出三个用户。

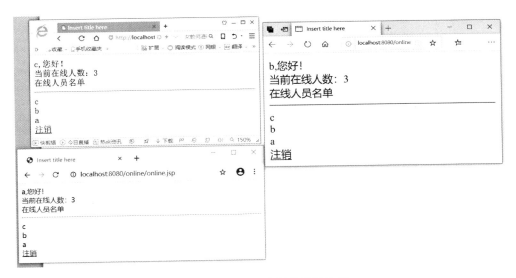
图 3.18　刷新三个浏览器的结果

接着,测试注销功能,先在第一个浏览器中点击"注销"按钮,并跳转到登录页面,在第二个浏览器就可以看到"当前在线人数:2"和"在线人员名单 c b",如图 3.19 所示,说明用户 a 已注销,并且在线人数和在线人员名单能够正确显示。同样,注销用户 b,也能正确显示,如图 3.20 所示。

图 3.19 用户 a 注销

图 3.20 用户 b 注销

习　　题

1. 编写一个注册表单,要求至少包含姓名、性别、所在城市、所读专业、兴趣爱好、个人简介等项目,根据需要使用文本框、复选框、单选按钮、下拉列表、多行文本框等,填写完成后返回显示所填信息的页面。

2. 编写一个登录页面,页面如图 3.21 所示,用户登录时可以设置是否"下次自动登录",使用 cookie 技术实现自动登录功能。

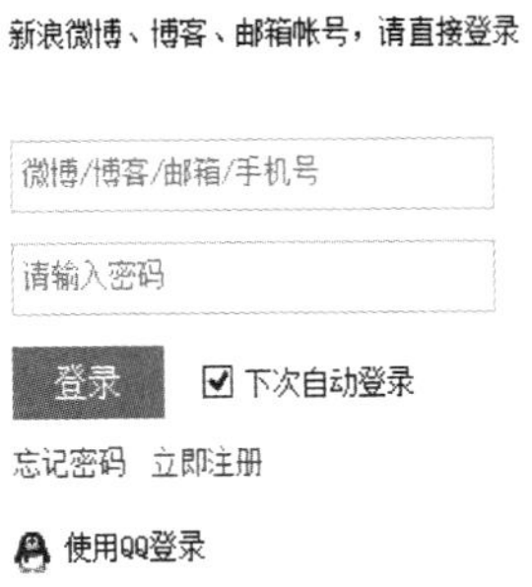

图 3.21　登录页面

3. 使用 session 实现简易的购物系统,用户登录之后选择商品,并放入购物车中。

第 4 章 Servlet 技术

【本章内容】
- 4.1 Servlet 概述
- 4.2 Servlet 编程
- 4.3 文件上传

4.1 Servlet 概 述

4.1.1 什么是 Servlet

Servlet 是运行在服务器端的 Java 程序,需要按照 Servlet 规范进行编写,用于处理客户端的请求。Servlet 技术在 Java EE 出现之前就已经存在了,在开发动态网页中得到广泛的应用,JSP 也是在 Servlet 的基础上发展起来的。在 MVC 设计模式中,Servlet 充当控制器。

4.1.2 Servlet 体系结构

Servlet 程序的运行需要 Servlet API 的支持,在 Tomcat 安装目录的 lib 目录中可以看到 servlet-api.jar 包。Servlet API 类主要放在 javax.servlet 和 javax.servlet.http 这两个包中,包的内容如图 4.1 所示。在 javax.servlet 包中定义了 Servlet 接口及相关的通用接口和类,在 javax.servlet.http 包中主要定义了与 HTTP 协议相关的接口和类,主要有 HttpServletRequest 接口、HttpServletResponse 接口、HttpServlet 类。

4.1.3 Servlet 常用的接口

①HttpServletRequest 接口:该接口是 request 内置对象的基类,request 对象是 HttpServletRequest 接口的一个实例,其实例化过程是自动的,无须自定义。

②HttpServletResponse 接口:该接口是 response 内置对象的基类,response 对象是 HttpServletResponse 接口的一个实例。

③HttpSession 接口:该接口是 session 内置对象的基类,session 对象是 HttpSession 接口的一个实例。在 Servlet 中,session 对象通过 request 对象的 getSession( )方法获得。

④ServletContext 接口:该接口是 application 内置对象的基类,application 对象是 ServletContext 接口的一个实例。在 Servlet 中 application 对象通过 HttpServletRequest 接口的 getServletContext( )方法获得。

图 4.1　javax.servlet 包和 javax.servlet.http 包的内容

4.2　Servlet 编程

4.2.1　Servlet 编程的流程

Servlet 编程的流程如图 4.2 所示，经历了创建 Servlet 类→配置 Servlet 类→部署 Servlet 类→调用 Servlet 类的过程，具体内容后面将进行详细的介绍。

4.2.2　创建 Servlet 程序

在进行 Web 应用开发时，由于使用的是 http 协议，所以创建的 Servlet 类是继承 Http Servlet类，HttpServlet 类包含很多方法，有 init( )初始化方法和 destroy( )销毁方法，也包含了 doGet( )、doPost( )和 Service( )。在 Eclipse 平台上创建 Servlet 具体步骤如下：

①在项目的 src 目录新建 Servlet 程序，右击 src 目录，选中"new→Servlet"，创建 Servlet 程序，如图 4.3 所示。

②如图 4.4 所示，在弹出的"Create Servlet"对话框中，填写包名和类名，通常 Servlet 程序要放在包中，这样方便调用，所创建的 Servlet 程序继承 javax.servlet.http.HttpServlet 类。设置包名为"demo"，类名为"ServletDemo"，然后点击"Next"按钮，进入下一步。

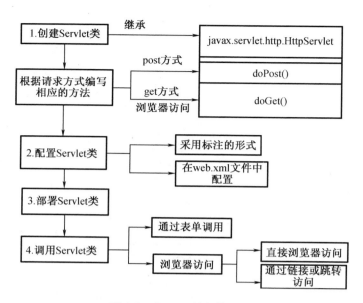

图 4.2　Servlet 编程的流程

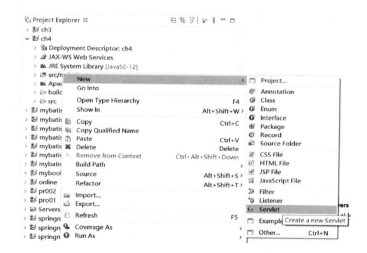

图 4.3　创建 Servlet 程序

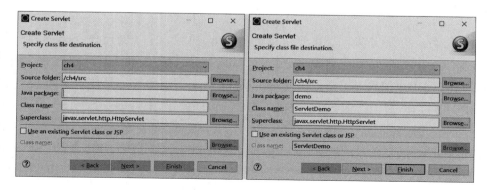

图 4.4　设置包名和类名

③设置 Servlet 程序的部署描述信息,如图 4.5 所示,系统已经自动设置了 URL mappings 名称"/ServletDemo",一般不做修改,使用默认的名称即可,然后点击"Next"按钮,进入下一步。

④最后,设置需要创建的方法,如图 4.6 所示,默认选中了"Constructors from superclass""Inherited abstract methods""doPost"和"doGet"方法,如果需要编写 init 和 destroy 方法,则在这两项前面打上钩,最后点击"Finish"按钮。

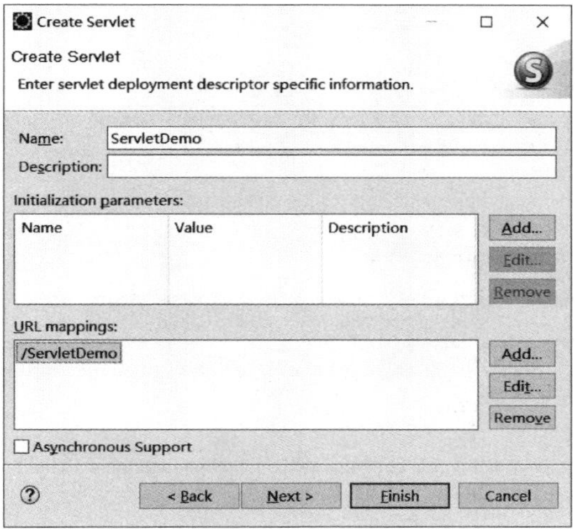

图 4.5　设置 Servlet 程序的部署描述信息 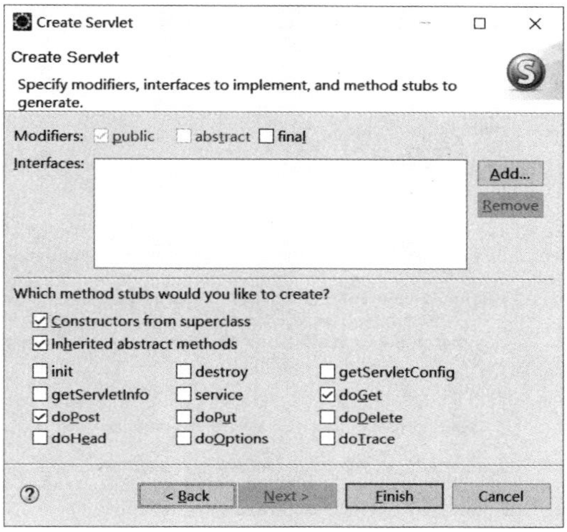　　　图 4.6　设置需要创建的方法

通过上面的步骤,创建了一个 Servlet 程序,是一个模板,如图 4.7 所示,要在相应的方法中写入实际的处理代码。这个程序包括了一个无参的构造方法 ServletDemo(),还有 doGet()和 doPost()。通过 Eclipse 创建 Servlet 程序,Servlet 的配置采用的是标注的形式,在图 4.7 中,代码"@WebServlet("/ServletDemo")"就是 Servlet 的标注,通过标注的形式进行配置,"/ServletDemo"就是这个 Servlet 的访问地址。

编写 doGet()方法,代码如下。

```java
protected void doGet(HttpServletRequest request, HttpServletResponse response)
throws ServletException, IOException {
    //TODO Auto-generated method stub
    response.setContentType("text/html;charset=utf-8");
    PrintWriter out = response.getWriter();
    out.println("正在在执行 doGet 方法");
}
```

4.2.3　运行 Servlet 程序

将项目部署到服务器,接着打开浏览器,输入"localhost:8080/ch4/ServletDemo",运行 ServletDemo 程序,结果如图 4.8 所示。

```
ServletDemo.java ⊠
 1 package demo;
 2
 3⊖import java.io.IOException;
 4 import javax.servlet.ServletException;
 5 import javax.servlet.annotation.WebServlet;
 6 import javax.servlet.http.HttpServlet;
 7 import javax.servlet.http.HttpServletRequest;
 8 import javax.servlet.http.HttpServletResponse;
 9
10⊖/**
11  * Servlet implementation class ServletDemo
12  */
13 @WebServlet("/ServletDemo")
14 public class ServletDemo extends HttpServlet {
15     private static final long serialVersionUID = 1L;
16
17⊖    /**
18      * @see HttpServlet#HttpServlet()
19      */
20⊖    public ServletDemo() {
21         super();
22         // TODO Auto-generated constructor stub
23     }
24
25⊖    /**
26      * @see HttpServlet#doGet(HttpServletRequest request, HttpServletResponse response)
27      */
28⊖    protected void doGet(HttpServletRequest request, HttpServletResponse response) throws ServletException, IOException {
29         // TODO Auto-generated method stub
30         response.getWriter().append("Served at: ").append(request.getContextPath());
31     }
32
33⊖    /**
34      * @see HttpServlet#doPost(HttpServletRequest request, HttpServletResponse response)
35      */
36⊖    protected void doPost(HttpServletRequest request, HttpServletResponse response) throws ServletException, IOException {
37         // TODO Auto-generated method stub
38         doGet(request, response);
39     }
40 }
```

图 4.7　Servlet 程序

图 4.8　ServletDemo 程序运行结果

4.2.4　Servlet 程序的生命周期

Servlet 程序部署在容器里，整个生命周期由容器进行管理。Servlet 的生命周期分为以下几个阶段：

1. Tomcat 容器装载 Servlet 类。
2. 创建 Servlet 实例。
3. 调用 Servlet 的 init( )方法，只有在第一次访问 Servlet 时才会执行 init( )方法。
4. 在访问 Servlet 时，根据客户端的请求方式执行相应的响应方法，有 doGet( )、doPost( )和 service( )，在方法中执行相应的处理。

5. 当出现以下三种情形之一时调用 Servlet 的 destroy( ) 方法,销毁该 Servlet 实例:

①Tomcat 重新启动;

②重载该项目;

③重新启动电脑。

【例 4.1】 编写一个 Servlet 程序,验证其生命周期。

在前面 Servlet 程序的基础上,完成 init( )、destroy( )、doGet( )和 doPost( )方法的编写,以验证 Servlet 程序的生命周期。

```java
public class ServletDemo extends HttpServlet {
    private static final long serialVersionUID = 1L;
    public ServletDemo() {
        super();
        //TODO Auto-generated constructor stub
    }
    public void init(ServletConfig config) throws ServletException {
        //TODO Auto-generated method stub
        System.out.println("正在初始化,生命周期开始");
    }
    public void destroy() {
        //TODO Auto-generated method stub
        System.out.println("正在销毁,生命周期结束");
    }
     protected void doGet (HttpServletRequest request, HttpServletResponse response) throws ServletException, IOException {
        //TODO Auto-generated method stub
        System.out.println("正在执行 doGet 方法");
        response.setContentType("text/html;charset=utf-8");
        PrintWriter out = response.getWriter();
        out.println("正在在执行 doGet 方法");
    }
     protected void doPost (HttpServletRequest request, HttpServletResponse response) throws ServletException, IOException {
        //TODO Auto-generated method stub
        System.out.println("正在执行 doPost 方法");
        response.setContentType("text/html;charset=utf-8");
        PrintWriter pw = response.getWriter();
        pw.println("正在在执行 doGet 方法");
    }
}
```

在本例中,采用在 web.xml 文件中配置 Servlet 的方法,具体配置代码如下。

```
<servlet>
    <servlet-name>ServletDemo</servlet-name>
    <servlet-class>demo.ServletDemo</servlet-class>
</servlet>
<servlet-mapping>
    <servlet-name>ServletDemo</servlet-name>
    <url-pattern>/ServletDemo</url-pattern>
</servlet-mapping>
```

其中，两个<servlet-name>名称要相同，这个名称是自定义的，可以跟Servlet类名不一样，<servlet-class>指定Servlet类所在的包和类名，<url-pattern>设置访问的URL，这个名称不需要跟实际的路径相同，可以跟<servlet-name>相同，也可以设置其他名称。访问Servlet程序可以输入访问的URL，假如<url-pattern>设置为"/ServletDemo1"，则访问时需要输入"localhost:8080/ch4/ServletDemo1"。

当把项目部署到服务器时，服务器将Servlet程序加载到内存，如果Servlet的配置有问题，服务器加载时就会报错。修改Servlet配置后必须重新部署到服务器才有效。

Servlet程序生命周期的验证如图4.9、图4.10所示。

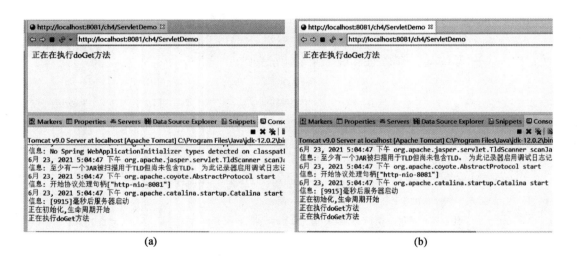

图4.9 Servlet程序生命周期的验证1

当在浏览器中输入访问地址"localhost:8080/ ch4/ ServletDemo"，第一次访问Servlet程序，会调用init()方法进行初始化，所以在控制台中可以看到相应的提示信息"正在初始化，生命周期开始"。init()方法只执行一次，通过地址栏进行访问，请求方式是get方式，因为执行了doGet()方法，所以看到提示信息"正在执行doGet方法"，地址使用的是创建Servlet时所设置的映射地址"/ ServletDemo"。

当再一次访问Servlet程序时，可以看到再次执行doGet()方法，输出提示信息"正在执行doGet方法"。当重载项目时，可以看到执行doDestroy()方法，输出提示信息"正在销毁，生命周期结束"。

```
<terminated> Tomcat v9.0 Server at localhost [Apache Tomcat] C:\Program Files\Java\jdk-12.0.2\bin\javaw.exe (
6月 23, 2021 5:04:47 下午 org.apache.catalina.startup.Catalina start
信息: [9915]毫秒后服务器启动
正在初始化,生命周期开始
正在执行doGet方法
正在执行doGet方法
6月 23, 2021 5:09:38 下午 org.apache.catalina.core.StandardServer await
信息: 通过关闭端口接收到有效的关闭命令。正在停止服务器实例。
6月 23, 2021 5:09:38 下午 org.apache.coyote.AbstractProtocol pause
信息: 暂停ProtocolHandler ["http-nio-8081"]
6月 23, 2021 5:09:41 下午 org.apache.catalina.core.StandardService stopInternal
信息: 正在停止服务[Catalina]
WARNING: An illegal reflective access operation has occurred
WARNING: Illegal reflective access by org.apache.catalina.loader.WebappClassLoaderBase (fi
WARNING: Please consider reporting this to the maintainers of org.apache.catalina.loader.W
WARNING: Use --illegal-access=warn to enable warnings of further illegal reflective access
WARNING: All illegal access operations will be denied in a future release
正在销毁,生命周期结束
6月 23, 2021 5:09:41 下午 org.apache.coyote.AbstractProtocol stop
信息: 正在停止ProtocolHandler ["http-nio-8081"]
6月 23, 2021 5:09:41 下午 org.apache.coyote.AbstractProtocol destroy
信息: 正在摧毁协议处理器 ["http-nio-8081"]
```

图 4.10　Servlet 程序生命周期的验证 2

至于 doPost( )方法的验证,由于这里没有涉及表单的提交,只是通过地址栏进行访问,这种请求方式是 get 方式,所以会调用 doGet( )方法。而当表单以 post 方式提交,这时就会调用 doPost( )方法。

4.3　文 件 上 传

一般文件上传需要借助第三方组件如 common-fileupload 和 smartUpload,而 Servlet 3.0 提供了很方便的文件上传功能,不需要借助第三方 jar 包。Servlet 3.0 以后的版本可以通过 HttpServletRequest 对象的 request.getPart( )、request.getParts( ) 获得 Part 接口,并在 Servlet 类前面加上注解@ MultipartConfig 实现文件上传,具体看下面的例子。

【例 4.2】　编写一个 Servlet 实现文件上传。

创建一个 upload.jsp 页面,表单提交方式设置为" post" enctype = " multipart/ form-data" 表示数据格式为二进制,表单元素 file 用于选择上传的文件,如果所选择的文件名包含有中文字符,那么 upload.jsp 页面声明 pageEncoding = " utf-8"。

```
<% @page language = "java" pageEncoding = "utf-8"% >
<!DOCTYPE html >
<html >
<head >
<meta http-equiv = "Content-Type" content = "text/html; charset = utf-8" >
<title >文件上传</title >
</head >
<body >
    <form action = "UploadServlet" method = "post" enctype = "multipart/ form-data" >
        请选择要上传的文件:<input type = "file" name = "file1"/ >
```

```
        <input type="submit" value="上传"/>
    </form>
</body>
</html>
```

创建一个 Servlet 程序 UploadServlet.java,在 Servlet 声明前加上@ MultipartConfig 注解,表示支持 HttpServletRequest 提供上传文件的功能,处理 multipart/form-data 类型的提交数据,通过 request 对象的 getPart()方法获得单一文件对象 part,再通过 part 对象的几个方法获得文件的相关信息。

getSize():获得文件大小。

getContentType():获得文件类型。

getHeader("Content-Disposition"):获得 Content-Disposition 头的信息,Content-Disposition: attachment; filename = "filename.xls",然后通过提取字符串获得上传的文件名。

getSubmittedFileName():支持 Tomcat 8.0 以上版本,用于获得上传的文件名。

write():将上传文件写入服务器的指定目录。

fileSizeThreshold:整数值设置,默认值为 0,若上传文件的大小超过了这个值,就会先写入缓存文件。

location:设置缓存文件的路径,默认值为空。

maxFileSize:限制文件上传大小,默认值为 -1L,表示不限制大小。

maxRequestSize:限制 multipart/form-data 请求个数,默认值为 -1L,表示不限制个数。

```java
@MultipartConfig
@WebServlet("/UploadServlet")
public class UploadServlet extends HttpServlet {
    protected void doPost(HttpServletRequest request, HttpServletResponse response) throws ServletException, IOException{
        //获取文件
        Part part = request.getPart("file1");
        //获取文件名,对于tomcat8.0以上版本,支持part.getSubmittedFileName(),
        String filename = part.getSubmittedFileName();
        request.setCharacterEncoding("utf-8");
        response.setContentType("text/html;charset=utf-8");
        PrintWriter out = response.getWriter();
        out.println("文件大小:"+part.getSize()+"<br>");
        out.println("文件类型:"+part.getContentType()+"<br>");
        out.println("文件名:"+filename+"<br>");
        //文件保存的路径
        String PathString = request.getServletContext().getRealPath("/upload");
        File filePath = new File(PathString);
        if(!filePath.exists()){
            filePath.mkdirs();    //创建目录
```

```
}
        out.println("文件保存的位置:" + "<br>");
        out.println(filePath + "<br>");
        //写入文件
        part.write(filePath + "/" + filename);
        out.println("已保存文件");
        }
}
```

运行结果如图 4.11 所示。

图 4.11 上传文件的运行结果

习　　题

利用 HttpServletRequest 的 getParts( )方法实现一次上传多个文件,并对文件类型和大小进行限制,限制只能上传 5 MB 以下的 JPG 格式的图片。

第 5 章　JDBC 数据库连接技术

【本章内容】
- 5.1 JDBC 概述
- 5.2 MySQL 数据库操作
- 5.3 数据库操作案例

5.1　JDBC 概 述

5.1.1　JDBC 简介

JDBC(Java Database Connectivity)是指 Java 数据库连接技术,是 Java 程序用于连接数据库的应用程序接口。使用 Java 语言编写的用于访问不同数据库的类和接口,支持 ANSI SQL-92 标准。通过调用这些类和接口所提供的方法,可以方便地连接各种不同的数据库,进而使用标准的 SQL 命令对数据库进行查询、插入、删除和更新等操作。

5.1.2　JDBC 的体系结构

JDBC 的体系结构包括四层,即 JDBC API、JDBC DriverManager、JDBC Driver API 和驱动程序,如图 5.1 所示。

①JDBC API:JDBC API 即 JDBC 应用程序接口,负责与 Java 应用程序进行交互,调用 JDBC 接口或类的方法,执行 SQL 语句并提取执行结果。

②JDBC DriverManager:JDBC 驱动程序管理器,管理各种驱动程序。

③JDBC Driver API:JDBC 驱动程序接口,这些接口是由 JDBC 规范提供的,由数据库厂商实现这些接口。

④驱动程序:包括四种不同类型的连接方式及所使用的驱动程序,具体在后面进行介绍。

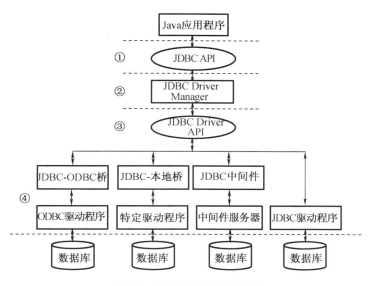

图 5.1　JDBC 的体系结构图

JDBC 连接数据库有四种驱动类型,各有不同的适用场景。

1. JDBC-ODBC 桥

这类驱动由 JDBC-ODBC 桥和一个 ODBC 驱动程序组成。JDBC-ODBC 桥采用 C 语言编写,用于将 JDBC API 访问指令转换为 ODBC API 指令,再借助计算机操作系统自带的数据源提供的 ODBC 驱动程序,完成对数据库的访问。

JDK 1.8 以上的版本已经不再支持 JDBC-ODBC 桥的类型,如果确实需要,比如要连接 Access 数据库,而 Access 数据库厂商并没有提供 JDBC 驱动程序,要借助 ODBC 驱动程序,这时就要采用 JDBC-ODBC 桥的方式,JDK 要使用 1.8 之前的版本。

2. JDBC 本地桥

这类驱动包括 JDBC 本地 API、本地的 C/C++ API 和数据库厂商提供的特定的驱动程序。JDBC 本地桥先将 JDBC API 访问指令转换为 C/C++ API 指令,C/C++ 指令使用特定的驱动程序操作数据库,每一台客户机都必须安装特定的驱动程序。C/C++ API 是针对特定数据库的接口,不同数据库的 API 是不同的,如 Oracle 调用接口(OCI)是用于本地调用 Oracle 驱动程序的,而 Oracle 公司也提供了针对 OCI 的 Oracle JDBC 驱动程序。

3. JDBC 中间件

这类驱动适用于在一个系统中需要访问多种数据库系统,使用不同驱动类型,这时需要一个中间件服务器。JDBC API 调用 JDBC 中间件,使用标准网络套接字与中间件服务器进行通信,套接字信息由中间件服务器转换成 DBMS 所需的调用格式,然后对相应的数据库服务器进行操作。

4. 纯 JDBC 驱动器

这类驱动是由数据库厂商使用 Java 语言遵循 JDBC 规范实现的驱动程序,是性能最高的驱动程序,使用 Java 语言通过套接字连接与数据库直接通信。这类驱动灵活,不需要在客户端或服务器上安装特殊的软件。

5.1.3 常用的类和接口

JDBC API 分为核心 API 和扩展 API：java.sql 包提供核心 API，提供基本的数据库操作；javax.sql 包提供扩展 API，支持连接池、数据源技术和分布式事务处理。两个包所包含的内容具体如图 5.2 所示。

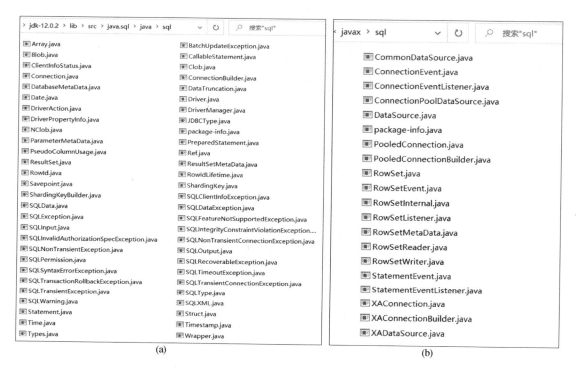

图 5.2 java.sql 包和 javax.sql 包的内容

1. java.sql.Driver 接口

这个接口由数据库厂商实现具体的数据库驱动程序类，驱动程序类的命名有一定的层次，如 MySQL 数据库的 JDBC 驱动程序类名为 com.mysql.jdbc.Driver，Oracle 数据库为 oracle.jdbc.driver.OracleDriver，Sybase 数据库为 com.sybase.jdbc2.jdbc.SybDriver。

2. java.sql.DriverManager 类

这个类用于管理多个驱动程序，能够根据设置的连接参数通过驱动程序与数据库建立连接。其主要的方法是 getConnection( )，用于创建连接对象。

第一种形式：getConnection(String url, String user, String password)，包含三个参数。

第二种形式：getConnection(String url)，只有一个参数，url 参数里面包含了账号和密码。

3. java.sql.Connection 接口

这个接口由数据库厂商在驱动程序中根据 JDBC 规范来创建连接，可实现应用程序和具体数据库的 socket 连接，这个过程对于用户是透明的。由 DriverManager 类的 getConnection( ) 方法得到 Connection 对象，Connection 对象常用的方法如下。

createStatement():创建语句处理对象 statement,用于发送 SQL 语句给数据库服务器,进而对数据库进行增删改查等操作。

prepareStatement():创建预处理语句对象 PreparedStatement,发送的 SQL 语句带有一个或多个参数,事先编译好,只要将参数直接传入编译过的语句执行代码中就会被执行,一条语句编译代码被缓存下来,多次执行,速度比 statement 速度快。同时,使用 PreparedStatement 预处理语句对象可以防止 SQL 注入,比较安全。

4. java.sql.Statement 接口

实现这个接口的对象将 SQL 语句提交到数据库,执行增删改查等操作,除了执行存储过程以外,一些派生的接口也接受参数,如 PreparedStatement 接口。其常用的方式有:

executeQuery(String sql):执行查询,结果储存在 ResultSet 对象中。

executeUpdate(String sql):执行增删改操作,返回一个整数,表示操作了多少条记录。

5. java.sql.ResultSet 接口

通过 statement 对象执行查询的结果得到 ResultSet 对象,数据保存的格式跟数据库表的格式一样,可以通过指针移动来获取记录,也可以通过迭代器和循环获取记录。

6. java.sql.SQLException 类

这个类用于处理在数据库操作过程发生的任何错误,开发人员可以根据捕获到的异常做进一步的处理。

5.1.4 JDBC 访问数据库的基本步骤

JDBC 访问数据库的基本步骤如下,代码以操作 MySQL 数据库为例。

1. 加载驱动程序

```
Class.forName("驱动程序名");
Class.forName("com.mysql.jdbc.Driver");           //MySQL5.6
Class.forName("com.mysql.cj.jdbc.Driver");        //MySQL8.0
```

2. 创建数据库连接对象

```
Connnection conn = DriverManager.getConnection(String url,String user,String password);
Connection conn = DriverManager.getConnection("jdbc:mysql://localhosts:3306/test","root","123456")
Connection conn = DriverManager.getConnection("jdbc:mysql://localhost:3306/test?user=root&password=123456");        //MySQL5.6
Connection conn = DriverManager.getConnection("jdbc:mysql://localhost:3306/test?useUnicode=true&characterEncoding=utf8&serverTimezone=Asia/Shanghai&user=root&password=123456");  //MySQL8.0
```

3. 创建语句处理对象

```
Statement stmt = conn.createStatement();
```

createStatement()方法是无参数的。

4. 指定 sql 语句

```
String sql = "sql 语句";
```

例如查询:`String sql = "select * from user";`

5. 执行查询或更新

如果是查询操作,则执行 executeQuery(sql),返回结果给记录集对象。

```
ResultSet rs = stmt.executeQuery(sql);
```

如果是添加、删除、更新等操作,则执行 executeUpdate(sql),没有返回结果。

```
int i = stmt.executeUpdate(sql);          //i 表示受影响的行数,也可以不返回
```

6. 关闭有关的资源

依照创建对象的相反顺序关闭资源,即先关闭 rs,再关闭 stmt,最后关闭 conn。

```
rs.close();
stmt.close();
conn.close();
```

5.2 MySQL 数据库操作

5.2.1 MySQL 命令

下面以留言板数据库系统为例,创建数据库的名称为 messageboard,创建一个管理员用户表 user。

在 MySQL 快捷菜单中,有一个菜单项"MySQL 8.0 Command Line Client",点击进入 MySQL 的命令窗口,然后输入 root 的密码,如图 5.3 所示。

(a)　　　　　　　　　　　　　(b)

图 5.3　MySQL 快捷菜单和命令窗口

1. 显示所有的数据库

```
mysql>show databases;
```

2. 创建数据库

```
mysql>create database messageboard;
```

3. 删除数据库、删除表

```
mysql>drop database books;
mysql>drop table book;
```

4. 打开数据库

mysql>use messageboard;

5. 创建表

mysql>create table user(
   id int(10) not null auto_increment,
   username varchar(20) not null,
   password varchar(20) not null,
   primary key(id));

创建表的命令如图 5.4 所示。

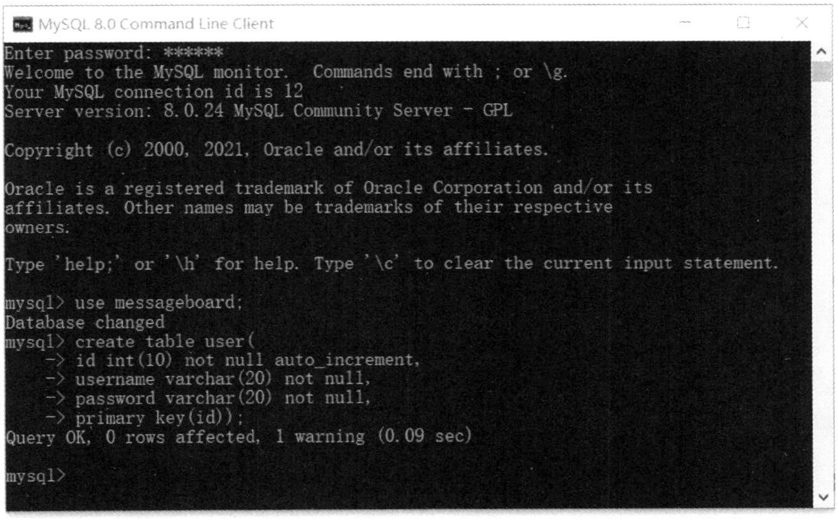

图 5.4 创建表的命令

6. 添加记录

mysql>insert into user(username,password)values("a","a");

7. 查询记录

mysql>select * from user;

5.2.2 Navicat 数据库管理工具

使用命令不太方便,我们可以借助一些数据库管理工具提高效率,如 MySQL 官方提供的 Workbench 工具。本书介绍另一款快速、可靠的数据库管理工具 Navicat,其中的一个版本 Navicat for MySQL 是专为 MySQL 设计的高性能数据库管理及开发工具,支持大部分 MySQL 最新版本的功能,包括触发器、存储过程、函数、事件、视图、管理用户等。

1. 创建连接

使用 Navicat 进行数据库管理,首先要创建连接,如图 5.5 所示,创建连接时,设置连接名如"myConnect",再设置 MySQL 数据库服务器的主机名或 IP 地址,在开发过程中,通常都是使用本机作为数据库服务器,所以填写为"localhost",端口为"3306",填入 root 账号的密码,这个密码是我们安装 MySQL 时设置的密码,要一致。填写好之后,点击"连接测试",如果弹出"连接成功",就表明连接创建好了。

图 5.5　创建连接

2. 创建数据库

创建数据库时,首先设置数据库名称,字符集选择"utf8—UTF-8 Unicode",这一项能够支持中文字符,并且排序规则选择"utf8_general_ci",如图 5.6 所示。

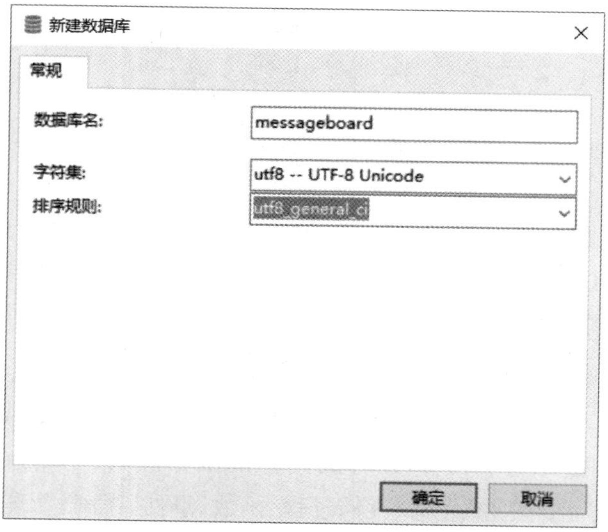

图 5.6　创建数据库

3. 创建表

创建表 user,设置 id、username 和 password 三个字段,类型和长度等信息如图 5.7 所示。同时,将字段 id 设置为"自动递增",这样添加一条记录时,id 会自动编号,不需要给它设置值。

图 5.7　创建表 user 的结构

4. 添加记录

添加三条记录,如图 5.8 所示。

图 5.8　添加三条记录

5.3　数据库操作实例

5.3.1　创建数据库

本实例使用前面创建的数据库 messageboard、表 user,创建一个 Web 项目 ch5,通过该项目演示如何使用 JDBC 连接数据库,进行增删改查等操作,项目的目录结构如图 5.9 所示。首先将在 MySQL 官网上下载的 JDBC 驱动包 mysql-connector-java-8.0.15.jar 拷到 lib 目录中,然后在 webapp 目录中创建 JSP 文件,每个 JSP 文件要设置 page 指令导入 sql 包。

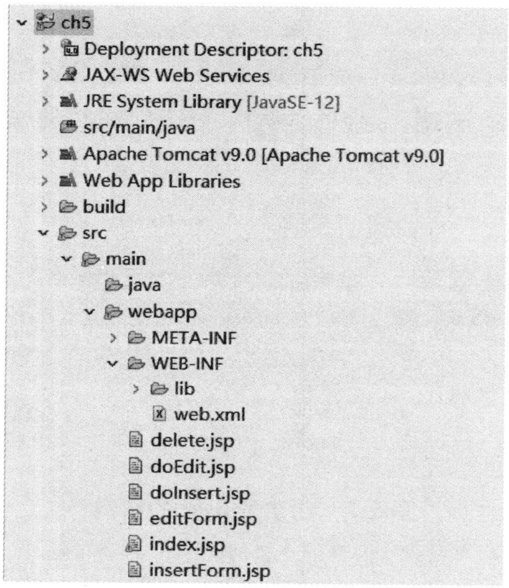

```
    ch5
    > Deployment Descriptor: ch5
    > JAX-WS Web Services
    > JRE System Library [JavaSE-12]
      src/main/java
    > Apache Tomcat v9.0 [Apache Tomcat v9.0]
    > Web App Libraries
    > build
    v src
      v main
          java
        v webapp
          > META-INF
          v WEB-INF
            > lib
              web.xml
          delete.jsp
          doEdit.jsp
          doInsert.jsp
          editForm.jsp
          index.jsp
          insertForm.jsp
```

图 5.9 项目的目录结构

1. 查询所有记录

index.jsp 文件：

```jsp
<%@page language="java" import="java.sql.*" pageEncoding="UTF-8"%>
<!DOCTYPE html>
<html>
  <head>
    <title>用户管理</title>
  </head>
  <body>
    <h3>显示所有记录</h3>
    <table border="1" width="500">
    <tr>
      <td>id</td>
      <td>用户名</td>
      <td>密码</td>
      <td>编辑</td>
      <td>删除</td>
    </tr>
    <%
      Class.forName("com.mysql.cj.jdbc.Driver");
      String url=" jdbc:mysql://localhost:3306/messageboard?useUnicode=true&characterEncoding=utf8&serverTimezone=Asia/Shanghai&user=root&password=123456";
      Connection conn=DriverManager.getConnection(url);
      Statement stmt=conn.createStatement();
      String sql="select * from user";
```

```jsp
ResultSet rs = stmt.executeQuery(sql);
while(rs.next()){
%>
<tr>
  <td><%=rs.getInt(1)%></td>
  <td><%=rs.getString(2)%></td>
  <td><%=rs.getString(3)%></td>
  <td><a href="editForm.jsp?id=<%=rs.getInt(1)%>">编辑</a></td>
  <td><a href="delete.jsp?id=<%=rs.getInt(1)%>">删除</a></td>
</tr>
<%
}
%>
</table>
<a href="insertForm.jsp">添加记录</a>
</body>
</html>
```

执行查询操作,结果保存在记录集中,记录集的结构跟数据表是一样的。记录集提供几个方法实现指针的移动操作和读取数据表字段值,如 rs.next() 用于判断是否存在下一条记录,并通过 while 循环逐条记录读取出来；rs.getInt(1) 表示取出第一个字段的值,因为第一个字段是整型,所以采用 rs.getInt() 方法；rs.getString(2) 表示取出第二个字段的值,第二个字段是字符型；rs.getString(3) 表示取出第三个字段的值。此外,还有 getDouble()、getLong()、getFloat()、getBoolean()、getDate()、getTime() 等,获取字段的值使用 get 加上相应的类型名,参数指明字段的序号。

运行结果如图 5.10 所示。

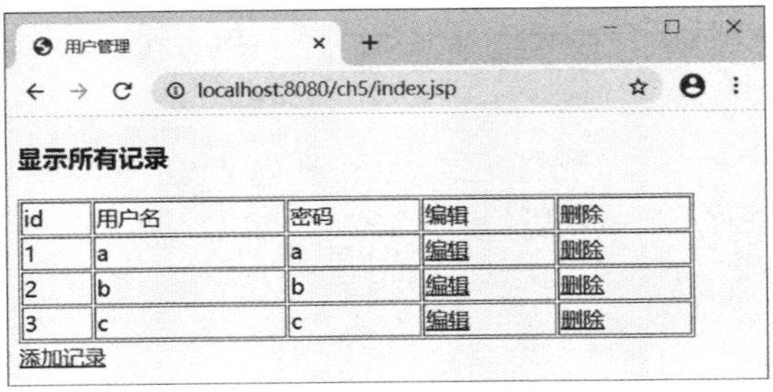

图 5.10 显示所有记录

2. 添加记录的页面

insertForm.jsp 文件:

```jsp
<body>
    <h3>添加记录</h3>
```

```
<form name = "form1" action = "doInsert.jsp" method = "post" >
用户名:< input type = "text" name = "username" > < br >
密码:< input type = "text" name = "password" > < br >
< input type = "submit" value = "添加" >
< input type = "reset" value = "重置" >
</form >
</body >
```

运行结果如图5.11所示。用户填写好新增记录的用户名和密码两个字段的值,点击"添加"按钮,则表单提交给 doInsert.jsp 程序进行处理。

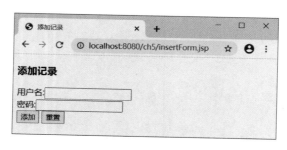

图5.11 添加记录的界面

3. 添加记录

doInsert.jsp 文件:

```
<body >
  <%
request.setCharacterEncoding("utf-8");
String username = request.getParameter("username");
String password = request.getParameter("password");
Class.forName("com.mysql.cj.jdbc.Driver");
String url = " jdbc:mysql://localhost:3306/messageboard? useUnicode = true char-
acterEncoding = utf8&serverTimezone = Asia/Shanghai&user = root&password =123456";
Connection conn = DriverManager.getConnection(url);
  String sql = "insert into user(username,password) values(?,?)";
  PreparedStatement pstmt = conn.prepareStatement(sql);
  pstmt.setString(1,username);
  pstmt.setString(2,password);
  pstmt.execute();
  out.println("已添加一条记录");
  response.setHeader("refresh","3;url = index.jsp");
%>
</body >
```

首先将表单数据的编码格式设置为utf-8,可避免出现中文乱码,然后定义 SQL 语句,在语句中使用了两个问号代表两个参数,再创建预处理语句对象 pstmt,对 SQL 语句预先编译,再通过 pstmt.setString(1,username)方法将第 1 个参数设置为表单元素 username 的值,pstmt.setString(2,password)方法将第 2 个参数设置为表单元素 password 的值,最后通过 pstmt.execute()方法将两个参数的值传入事先编译好的代码中执行,这样就将表单元素 username 和 password 的值作为字段 username 和 password 的值插入到 user 表中,而字段 id 是自动编号,每添加一条记录,会自动加上编号。

运行结果如图 5.12 所示,提示"已添加一条记录"。

图 5.12　成功添加记录界面

4. 删除记录

delete.jsp 文件:

```
<%@ page language = "java" import = "java.sql.*" pageEncoding = "UTF-8"%>
<%
    String id = request.getParameter("id");
    Class.forName("com.mysql.cj.jdbc.Driver");
    String url = "jdbc:mysql://localhost:3306/messageboard?useUnicode = true&characterEncoding = utf8&serverTimezone = Asia/Shanghai&user = root&password = 123456";
    Connection conn = DriverManager.getConnection(url);
    String sql = "delete from user where id = ?";
    PreparedStatement pstmt = conn.prepareStatement(sql);
    pstmt.setInt(1,Integer.parseInt(id));
    pstmt.execute();
    out.println("已删除指定的记录");
    response.setHeader("refresh","3;url = index.jsp");
%>
```

删除记录是通过在 index.jsp 页面中,将要删除的那条记录的 id 作为参数通过地址栏传递完成操作的。在 delete.jsp 程序中,首先利用 request.getParameter("id")读取到这个 id,然后通过 where 子句设置删除指定 id 的记录。

运行结果如图 5.13 所示,显示"已删除指定的记录",然后通过 response.setHeader()设置 refresh 响应头,实现隔 3 s 自动跳转到 index.jsp 页面。

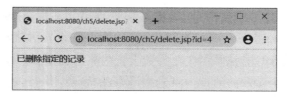

图 5.13 删除记录界面

5. 更新记录的页面

editForm.jsp 文件：

```jsp
<body>
    <h3>更新记录</h3>
    <%
    String id = request.getParameter("id");
    session.setAttribute("id",id);
    Class.forName("com.mysql.cj.jdbc.Driver");
        String url = "jdbc:mysql://localhost:3306/messageboard?useUnicode=true characterEncoding=utf8&serverTimezone=Asia/Shanghai&user=root&password=123456";
        Connection conn = DriverManager.getConnection(url);
        String sql = "select * from user where id=?";
        PreparedStatement pstmt = conn.prepareStatement(sql);
    pstmt.setInt(1,Integer.parseInt(id));
    ResultSet rs = pstmt.executeQuery();
    rs.next();
    %>
    <form name="form1" action="doEdit.jsp" method="post">
    id:<input type="text" name="id" value="<%=rs.getInt(1)%>" readonly>
<br>
    用户:<input type="text" name="username" value="<%=rs.getString(2)%>">
<br>
    密码:<input type="text" name="password" value="<%=rs.getString(3)%>">
<br>
    <input type="submit" value="更新">
    <input type="reset" value="重置">
    </form>
</body>
```

更新记录的操作比较复杂，具体分为两步：首先将指定的记录在表单中显示出来，在 index.jsp 页面中，String id = request.getParameter("id") 语句将要更新的记录的 id 作为参数传递到更新页面 editForm.jsp 文件中；然后通过 session.setAttribute("id",id)，将 id 保存为 session 属性，这样就可以将 id 的值由 index.jsp 页面传递到第三个页面 doEdit.jsp。

更新记录运行结果如图5.14所示,可以看到,指定id的那条记录各个字段的值已显示在表单相应的位置,id已设置为只读,不能修改,其他两个字段可以修改,然后单击"更新"完成操作。

图5.14 更新记录界面

6. 更新记录

doEdit.jsp文件:

```jsp
<%@page language="java" import="java.sql.*" pageEncoding="UTF-8"%>
<%
    String id = (String)session.getAttribute("id");
    String username = request.getParameter("username");
    String password = request.getParameter("password");
    Class.forName("com.mysql.cj.jdbc.Driver");
    String url = "jdbc:mysql://localhost:3306/messageboard?useUnicode=true&characterEncoding=utf8&serverTimezone=Asia/Shanghai&user=root&password=123456";
    Connection conn = DriverManager.getConnection(url);
    String sql = "update user set username=?,password=? where id=?";
    PreparedStatement pstmt = conn.prepareStatement(sql);
    pstmt.setString(1,username);
    pstmt.setString(2,password);
    pstmt.setInt(3,Integer.parseInt(id));
    pstmt.execute();
    out.println("已更新指定的记录");
    response.setHeader("refresh","3;url=index.jsp");
%>
```

更新记录是将修改后的表单数据更新到指定的记录中。首先通过 String id=(String)session.getAttribute("id")读取出id,然后连接数据库,将表单中各个元素的值更新到指定的记录,运行结果如图5.15所示,显示"已更新指定的记录",并在3 s后自动跳转到index.jsp,从index.jsp页面可以看到已更新的结果,如图5.16所示。

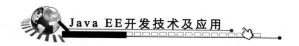

图 5.15 已更新记录的提示信息

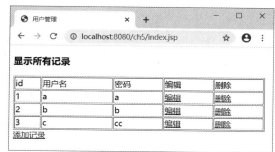

图 5.16 更新后的 index.jsp

第6章 JavaBean 组件技术

【本章内容】

- 6.1 JavaBean 概述
- 6.2 JavaBean 的生命周期
- 6.3 JavaBean 的应用

6.1 JavaBean 概述

6.1.1 什么是 JavaBean

JavaBean 是 Java 语言的一种组件技术,通过 JavaBean 可以实现代码的复用。可以将实现一定功能的代码封装为一个 JavaBean,这样就可以被其他程序调用,以达到代码复用的目的。

一个 JavaBean 就是一个 Java 类,它的编写具有一定的规范,具体如下:

(1)是一个公开的类,即 public class 类名,类名和 JavaBean 文件名相同,通常要放在一个包里面;

(2)必须有一个无参的构造方法,在构造方法中可以对属性进行初始化;

(3)如果需要与外部程序进行数据交互,需要设置属性,属性需要设置为私有的,通过 set×××()和 get×××()方法来让外部程序获取和设置 JavaBean 属性值,所定义的方法必须是公开的,set×××()和 get×××()方法名中属性名部分第一个字母要大写,如 getUserName();

(4)当需要实现序列化时,通过实现 java.io.Serialixable 或 java.io.Externalixable 接口以支持序列化。

【例6.1】 一个 JavaBean 示例。

JavaBeanDemo.java:

```
package javabean;
public class JavaBeanDemo{
  private String username;
  private String password;
  public JavaBeanDemo(){
    System.out.println("JavaBean 正在初始化……");
    this.username = "john";
```

```
    this.password = "abc";
  }
  public void setUsername(String username){
    this.username = username;
  }
  public String getUsername(){
    return this.username;
  }
  public void setPassword(String password){
    this.password = password;
  }
  public String getPassword(){
    return this.password;
  }
}
```

6.1.2 在 JSP 中调用 JavaBean

在 JSP 中调用 JavaBean 有两种方式,一种方式是使用 new 来创建 JavaBean 类的实例对象,然后进行属性设置或调用其定义的方法。采用这种方式要先导入 JavaBean 类所在的包。

```
<%
  JavaBeanDemo jbDemo = new JavaBeanDemo();
  jbDemo.setUsername = "liming";
  jbDemo.setPassword = "123456";
%>
```

另一种方式是使用 useBean 动作标签 <jsp:useBean>,这时可以通过 <jsp:setProperty> 标签设置 JavaBean 的属性,可利用 <jsp:getProperty> 标签获取 JavaBean 的属性。

```
<%@page language="java" pageEncoding="utf-8"%>
<html>
<head>
<title></title>
</head>
<body>
<!-- 下面调用 JavaBeanDemo,并设置其属性 -->
  <jsp:useBean id="jbDemo" class="javabean.JavaBeanDemo" scope="request">
</jsp:useBean>
<jsp:setProperty name="jbDemo" property="username" value="liming"/>
<jsp:setProperty name="jbDemo" property="password" value="123456"/>
</body>
```

6.2 JavaBean 的生命周期

使用 new 创建 JavaBean 类的实例对象,只在当前页面有效,而使用 < jsp:useBean > 调用 JavaBean 有四个作用范围,即 page、request、session 和 application。下面创建 useJavaBeanRequest1.jsp 和 useJavaBeanRequest2.jsp。

useJavaBeanRequest1.jsp:

```
<%@page language = "java" contentType = "text/html; charset = UTF-8" pageEncoding = "UTF-8"%>
<!DOCTYPE html>
<html>
<head>
<meta charset = "UTF-8">
<title>Insert title here</title>
</head>
<body>
<!-- 下面调用 JavaBean,并设置其属性 -->
  <jsp:useBean id = "jbDemo" class = "javabean.JavaBeanDemo" scope = "request">
</jsp:useBean>
<jsp:setProperty name = "jbDemo" property = "username" value = "liming"/>
<jsp:setProperty name = "jbDemo" property = "password" value = "123456"/>
<jsp:forward page = "useJavaBeanRequest2.jsp"/>
</body>
</html>
```

useBeanRequest2.jsp:

```
<%@page language = "java" contentType = "text/html; charset = UTF-8" pageEncoding = "UTF-8"%>
<!DOCTYPE html>
<html>
<head><title></title></head>
<body>
<!-- 取得 Javabean 属性 -->
  <jsp:useBean id = "jbDemo" class = "javabean.JavaBeanDemo" scope = "request">
</jsp:useBean>
 username:<jsp:getProperty name = "jbDemo" property = "username"/><br>
 password:<jsp:getProperty name = "jbDemo" property = "password"/>
</body>
```

运行 useJavaBeanRequest1.jsp,结果如图 6.1 所示。

图 6.1 useJavaBeanRequest1.jsp 的运行结果

运行 useJavaBeanRequest1.jsp 时，首先创建 JavaBean 的实例对象 jbDemo，创建时会执行 JavaBeanDemo()无参的构造方法进行初始化，所以在控制台输出"JavaBean 正在初始化……"。接下来设置 jdDemo 对象的两个属性的值，再通过 <jsp:forward> 实现服务器端的跳转，直接转向执行 useJavaBeanRequest2.jsp 程序，因为没有重新发出一个请求，没有再重新创建 JavaBean 的实例对象 jbDemo，所以控制台没有再输出"JavaBean 正在初始化……"，在 useJavaBeanRequest1.jsp 程序中创建的 jbDemo 对象仍然有效，输出的 username 和 password 属性的值是在 useBeanRequest1.jsp 中设置的值，并不是初始化的初值。

对于其他三个作用域 page、session 和 application，大家可以尝试一下，体会一下它们各自的作用范围。

6.3 JavaBean 的应用

下面通过一个例子来介绍怎么对复用代码进行封装，即如何将实现工资计税的代码封装为一个 JavaBean。在编写的时候，所要考虑的是需不需要设置属性，如果需要跟外部程序进行数据交互，这时就要设置属性，否则，就不需要设置属性。

个人所得税的计算公式是：个人所得税 = 工资中应征税部分 × 税率，税率是分级进行计算的，税率在一个时期内计算方法是固定的，所以计算个人所得税应该先取得工资的值，再进行计算。

6.3.1 方法一

设置工资作为 TaxRatioBean 的属性，为调用它的程序提供数据。

TaxRatioBean.java(有属性的):

```java
package javabean;
public class TaxRatioBean{
    private double salary;//工资
    public void setSalary(double salary){
      this.salary = salary;
    }
    Public double getSalary(){
      Return this.salary;
    }
    public float getTaxRatio(){
      float tax;//应缴纳的所得税
      float start =3500;//需要缴税的起点
      float taxSalary = salary - start;//工资中需要缴税的部分
      if(taxSalary < =0)
        {
         tax =0;
        }
      else if(taxSalary <500){
         tax = taxSalary * 5/100;
      }else if(taxSalary <2000){
         tax = taxSalary * 10/100 -25;
      }else if(taxSalary <5000){
         tax = taxSalary * 15/100 -125;
      }else{
         tax = taxSalary * 20/100 -375;
      }
return tax;
    }
}
```

taxForm.jsp:

```jsp
< body >
 < form method = "post" action = "doTax.jsp" >
 请输入您的工资 < input type = "text" name = "salary" />
 < input type = "submit" value = "计算所得税" />
  < /form >
< /body >
```

doTax.jsp：

```jsp
<body>
<%
    String salary = request.getParameter("salary");
    float salary1 = Float.parseFloat(salary);
    TaxRatioBean taxB = new TaxRatioBean();
    taxB.setSalary(salary1);
    float tax = taxB.getTaxRatio(); //调用了TaxRatioBean计算税收
    out.print("工资是:" + salary + " 应缴纳的税是:" + tax);
%>
</body>
```

6.3.2 方法二

工资通过 TaxRatioBean 中的 getTaxRatio( ) 方法的参数进行传递，而不设置为 TaxRatioBean 的属性。

TaxRatioBean.java（无属性的）：

```java
package javabean;
public class TaxRatioBean{
    public float getTaxRatio(float salary){
        float tax;//应交的税
        float start = 3500;//需要缴税的起点
        float taxSalary = salary-start;//工资中需要缴税的部分
        if(taxSalary <= 0)
        {
            tax = 0;
        }
        else if(taxSalary < 500){
            tax = taxSalary * 5 / 100;
        }else if(taxSalary < 2000){
            tax = taxSalary * 10 / 100 - 25;
        }else if(taxSalary < 5000){
            tax = taxSalary * 15 / 100 - 125;
        }else{
            tax = taxSalary * 20 / 100 - 375;
        }
        return tax;
    }
}
```

taxForm.jsp：

```
<body>
<form method="post" action="doTax.jsp">
请输入您的工资<input type="text" name="salary"/>
<input type="submit" value="计算所得税"/>
</form>
</body>
```

doTax.jsp：

```
<body>
  <%
   String salary = request.getParameter("salary");
   float salary1 = Float.parseFloat(salary);
   TaxRatioBean taxB = new TaxRatioBean();
   float tax = taxB.getTaxRatio(salary1); //调用了TaxRatioBean计算税收
   out.print("工资是:" + salary + " 应缴纳的税是:" + tax);
  %>
</body>
```

JavaBean在数据库操作中主要是用于封装数据模型和实现数据库操作的业务逻辑，具体可以参考后面8.3.3节模型编写部分的代码。

习　　题

1. 创建一个JavaBean，用于实现学生成绩的等级评定。编写一个JSP页面，当输入学生成绩时调用JavaBean，给出学生的等级评定。(要求编写两种形式的JavaBean，一种是没有设置属性的，另一种是设置属性的。 <60:不及格；<70:及格；<80:中等；<90:良好;其他:优秀)

2. 创建一个JavaBean，实现输入数字和字符的验证码功能。

第 7 章　Java EE 软件架构模式

【本章内容】

- 7.1　Model1 架构模式
- 7.2　MVC 架构模式
- 7.3　多层架构模式
- 7.4　EL 表达式和 JSTL 标准标签库

软件架构是指在一定的设计原则基础上,从不同角度对组成系统的各部分进行搭配和安排,形成系统的多个结构而组成的架构。软件架构包括该系统的各个组件、组件的外部可见属性及组件之间的相互关系。组件的外部可见属性是指其他组件对该组件所做的假设。

软件架构模式很多时候被认为是设计模式,实际上它跟设计模式是不同的。Java 语言提供 23 种设计模式,包括单例模式、工厂方法模式、原型模式、代理模式等。软件架构模式和设计模式虽然都提供一套可重用的设计方法,但二者具有不同的粒度,相较软件架构模式,设计模式更偏重于定义出某个功能组件的微观结构。

采用 Java EE 技术开发的软件,从简单到复杂,主要有以下这三种架构模式。

7.1　Model1 架构模式

经典的 Model1 架构模式的工作流程如图 7.1 所示,整个系统都是由 JSP 程序组成的,JSP 程序一方面要负责页面的显示,另一方面还要实现业务逻辑,并控制流程,导致 JSP 程序中夹杂着大量的 Java 代码,这样不便于系统维护和代码的复用。Model1 架构模式比较简单,实现比较快速,所以比较适合于小型的应用系统。

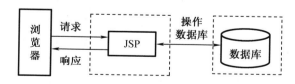

图 7.1　经典的 Model1 架构模式的工作流程

在经典的 Model1 架构模式的基础上,将实现业务逻辑的代码独立出来,成为 JavaBean 组件,这种改进的 Model1 架构模式如图 7.2 所示。第 5 章的数据库操作实例采用的就是经典的 Model1 架构模式,在系统中存在重复的连接数据库的代码,如果数据库链接地址要改变,这时需要在多个地方进行修改。而在图 7.2 所示的这种架构模式中,将对数据库的操作独立为 JavaBean 组件，大大地减少了重复代码，也更易于维护。

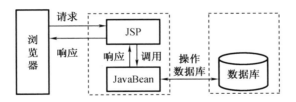

图 7.2　改进的 Model1 架构模式

下面以实现用户登录功能为例，分析一下经典 Model1 模式和改进的 Model1 模式的构成。

7.1.1　经典 Model1 模式

经典的 Model1 模式实现用户登录由 4 个文件组成，即登录页面 login.jsp、处理登录的 doLogin.jsp 文件、登录成功的 wel.jsp 文件和登录失败的 fail.jsp 文件。

登录页面 login.jsp：

```
<body>
 <form action="doLogin.jsp" method="post">
  username:<input type="text" name="username"><br>
  password:<input type="password" name="password"><br>
  <input type="submit" value="login">
 </form>
</body>
```

处理登录的程序 doLogin.jsp：

```
<%
 String username = request.getParameter("username");
 String password = request.getParameter("password");
 if(!username.equals("")&&!password.equals("")){
   Class.forName("com.mysql.cj.jdbc.Driver");
   Connection conn = DriverManager.getConnection("jdbc:mysql://localhost:3306/test?useUnicode=true&characterEncoding=utf8&useSSL=false&serverTimezone=Asia/Shanghai","root","123456");
   Statement stmt = conn.createStatement();
   String sql = "select * from user where username = '"+username+"' and password = '"+password+"'";
   ResultSet rs = stmt.executeQuery(sql);
   if(rs.next()){
     System.out.println("登录成功!");
     request.getRequestDispatcher("wel.jsp").forward(request,response);
   }else{
     System.out.println("登录失败!");
```

```
      request.getRequestDispatcher("fail.jsp").forward(request,response);
     }
   }else{
    out.println("用户名和密码都不能为空,请重新输入!");
   }
%>
```

登录成功的 wel.jsp 文件:

```
<body>
    <h3>你已成功登录页面!</h3>
</body>
```

登录失败的 fail.jsp 文件:

```
<body>
    <h3>登录失败!</h3>
</body>
```

7.1.2 改进的 Model1 模式

改进的 Model1 模式实现用户登录由 4 个 jsp 文件和 1 个 JavaBean 文件组成,即登录页面 login.jsp、处理登录的 doLogin.jsp 文件、实现登录验证的 JavaBean 组件 LoginCheckBean.java、登录成功的 wel.jsp 文件和登录失败的 fail.jsp 文件。

登录页面 login.jsp:

```
<body>
  <form action="doLogin.jsp" method="post">
    username:<input type="text" name="username"><br>
    password:<input type="password" name="password"><br>
    <input type="submit" value="login">
  </form>
</body>
```

处理登录的组件 LoginCheckBean.java:

```
Package javabean
public class LoginCheckBean{
private String username;
  private String password;
  <!--省略 setter 和 gettter 方法-->
  public boolean isLogin(){
    Class.forName("com.mysql.cj.jdbc.Driver");
    Connection conn = DriverManager.getConnection("jdbc:mysql://localhost:3306/test?useUnicode=true&characterEncoding=utf8&useSSL=false&serverTimezone=Asia/Shanghai","root","123456");
    Statement stmt = conn.createStatement();
    String sql = "select * from user where username='"+this.username+"' and password='"+this.password+"'";
```

```
    ResultSet  rs = stmt.executeQuery(sql);
    if(rs.next()){
      return true;
    }else{
      Return false;
    }
}
```

处理登录的程序 doLogin.jsp：

```
<%@page language="java" import="java.sql.*" contentType="text/html;charset=UTF-8"
 pageEncoding="UTF-8" import="javabean.LoginCheckBean"%>
<%
 String username = request.getParameter("username");
 String password = request.getParameter("password");
 LoginCheckBean  lcBean = new LoginCheckBean();
 lcBean.setUsername(username);
 lcBean.setPassword(password);
 if(lcBean.isLogin()){
   System.out.println("登录成功!");
   request.getRequestDispatcher("wel.jsp").forward(request,response);
 }else{
   System.out.println("登录失败!");
   request.getRequestDispatcher("fail.jsp").forward(request,response);
 }
%>
```

wel.jsp 文件和 fail.jsp 文件与经典的 Modell 模式中的文件相同。

从上面的例子可以看出，连接数据库进行验证的代码已经独立出来成为一个 JavaBean 组件，但流程控制仍然由 JSP 程序的 Java 代码实现。

7.2 MVC 架构模式

MVC 即 Model-View-Controller（模型－视图－控制器），是当前 Java EE 平台主流的架构模式。它将应用程序分解为模型、视图和控制器，使得应用程序结构更加清晰，也有利于代码的复用，有利于团队开发。

模型、视图和控制器这三部分在应用程序中的分工如下。

模型：负责实现应用程序的业务逻辑，封装各种数据及对数据的各种处理方法，然后返回结果。

视图：负责应用程序对用户的显示，从用户那里获取输入数据并通过控制器传给模型处理，然后再通过控制器将模型返回的结果显示给用户。

控制器：负责控制应用程序的流程，它相当于中介的作用，一方面接收从视图传过来的数据，另一方面调用模型的某个业务来进行处理，并根据模型返回的结果选择视图显示结果。

MVC 架构模式的工作流程如图 7.3 所示，具体通过下面的实例进行分析。

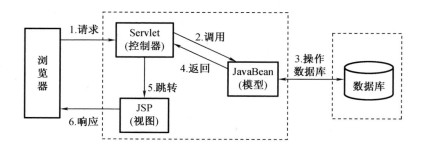

图 7.3　MVC 架构模式的工作流程

采用 MVC 设计模式实现登录需要 5 个文件，即 3 个 JSP 文件(login.jsp、wel.jsp 和 fail.jsp)、1 个 JavaBean 文件(LoginCheckBean.java)和 1 个 Servlet 程序(LoginServlet.java)。

LoginCheckBean.java 是实现登录验证的 JavaBean，这里没有连接数据库进行验证，只是根据给定的一个用户名"liming"和密码"123456"进行验证。

LoginCheckBean.java：

```java
package mvc;
public class LoginCheckBean {
private String username;
  private String password;
  <- -省略 setter 和 gettter 方法 - - >
  public boolean isLogin(){
    if(username.equals("liming")&&password.equals("123456")){
      return true;
    }else{
      return false;
    }
  }
}
```

LoginServlet.java 是由 Servlet 程序充当控制器的，一方面接收客户端提交的数据 username 和 password，另一方面调用登录验证 LoginCheckBean，根据 LoginCheckBean 的执行结果来控制流程，决定跳转到哪个 JSP 文件。

LoginServlet.java：

```java
package mvc;
import java.io.IOException;
import java.io.PrintWriter;
```

```
import javax.servlet.ServletException;
import javax.servlet.http.HttpServlet;
import javax.servlet.http.HttpServletRequest;
import javax.servlet.http.HttpServletResponse;
import javax.servlet.http.HttpSession;
@Servlet("/LoginServlet")
public class LoginServlet extends HttpServlet{
  public void doPost(HttpServletRequest request,HttpServletResponse response)
throws ServletException,IOException{
  response.setContentType("text/html");
  String username=request.getParameter("username");
  String password=request.getParameter("password");
  loginCheckBean lcBean=new loginCheckBean();
  lcBean.setUsername(username);
  lcBean.setPassword(password);
  if(lcBean.isLogin()){
    request.getRequestDispatcher("wel.jsp").forward(request,response);
  }else{
    request.getRequestDispatcher("fail.jsp").forward(request,response);
  }
 }
}
login.jsp
```

登录页面 login.jsp：

```
<%@page language="java" import="java.util.*" pageEncoding="utf-8"%>
<!DOCTYPE html>
<html>
  <head>
    <title>login 页面</title>
  </head>
  <body>
  <form action="LoginServlet" method="post">
    username:<input type="text" name="username"><br>
    password:<input type="password" name="password"><br>
    <input type="submit" value="login">
  </form>
  </body>
</html>
```

登录成功页面 wel.jsp：

```
<body>
  login success!
</body>
```

登录失败页面 fail.jsp：

```
<body>
    login failure!
</body>
```

7.3 多层架构模式

在企业级开发中,由于应用系统的业务逻辑比较复杂,所以需要进行分解,实现"高内聚,低耦合",使开发人员分工更明确,将精力更专注于应用系统核心业务逻辑的分析、设计和开发,加快项目的进度,提高开发效率,也有利于项目的更新和维护工作,因此 MVC 架构模式逐渐演变为多层架构模式。目前,很多主流的框架组合实现的就是多层架构模式,如SSH(Spring + Struts2 + Hibernate)和 SSM(Spring + SpringMVC + Mabatis)。

多层架构模式如图 7.4 所示,从图中可以看到,相比较 MVC 模式,多层架构模式将MVC 模式中的模型分解为三个部分,即持久化类、数据访问类和业务逻辑类,原来的视图和控制器构成表现层,形成了表现层、业务逻辑层、数据访问层和数据库层的多层架构。

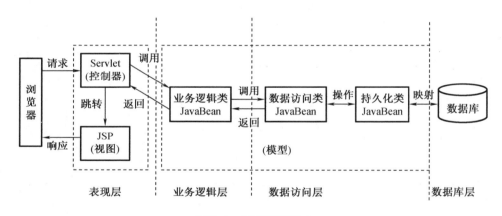

图 7.4　多层架构模式

7.4　EL 表达式和 JSTL 标准标签库

在 MVC 和多层架构模式下,系统已经被分解为模型、视图和控制器三个独立的部分,JSP 文件只是控制界面的显示,不再进行业务逻辑和流程的控制,因此在 JSP 中应尽量减少Java 代码段,那么对于控制器传递过来的数据,就需要 EL 和 JSTL 配合使用来展示数据。

7.4.1　EL 表达式

EL(Express Lanuage),即表达式语言,从 JSP 2.0 开始被引入,用于在 JSP 中获取数据,以减少 JSP 代码段。EL 只能用于读取 JSP 中各种对象的数据,而不能进行设置。其语法格式为：${表达式}。

1. EL 提供的对象

在 JSP 中，有内置对象 page、request、session 和 application，可以通过设置这些对象的属性值的方式保存数据，然后通过读取属性值得到这些数据，也可以通过 cookie 对象保存数据。JavaBean 也可以设置属性值、读取属性值、读取表单元素的值及应用程序初始化参数的值；通过 pageContext 对象获取页面上的内置对象，由这些内置对象提供的方法获取相应的信息；通过 request 对象获取请求头参数的值。EL 提供了相应的对象可以获取到这些数据，表 7.1 列出了 EL 提供的对象及对应的底层方法和作用。

表 7.1 EL 提供的对象及对应的底层方法和作用

EL 对象	对应的底层方法	作用
pageScope. 属性名 ${pageScope.name}	pageContext. getAttribute("属性名") pageContext.getAttribute("name")	获取 page 范围内属性的值
requestScope. 属性名 ${requesetScope.name}	request. getAttribute("属性名") request.getAttribute("name")	获取 request 范围内属性的值
sessionScope. 属性名 ${sessionScope.name}	session. getAttribute("属性名") session.getAttribute("name")	获取 session 范围内属性的值
applicationScope. 属性名 ${applicationScope.name}	application. getAttribute("属性名") application.getAttribute("name")	获取 application 范围内属性的值
param. 参数名 ${param.password}	request. getParameter("表单元素名") request.getParameter("password")	获取指定表单元素的值
paramValues. 参数名 ${paramValues.hobby}	request. getParameterValues("表单元素名") request.getParameterValues("hobby")	获取指定表单元素的多个值
initParam. 参数名 ${initParam.username}	ServletContext. getInitParameter("参数名") ServletContext.getInitParameter("username")	获取应用程序的初始化参数
cookie. 键名 ${cookie.name}	cookie 对象名. getValue("键名") cookie1.getValue("name")	获取 cookie 对象指定键名的值
JavaBean 对象名. 属性名 ${user.username}	JavaBean 对象名. get 属性名 user.getUsername()	获取 JavaBean 对象指定属性的值
header[报头名称] ${header["user-agent"]}	request. getHeader("报头名称") request.getHeader("user-agent")	获取指定请求报头名称的值
pageContext. JSP 内置对象. 方法名 ${pageContext.request.contextPath}	JSP 内置对象. get 方法名() request.getContextPath()	获取指定内置对象的 get 方法的值

表 7.1 列出的这些 EL 表达式获取数据的方式，在后面章节中会有相应的应用，这里就不具体展开介绍。

2. EL 读取 map 类型的数据

EL 可以读取 map 类型的数据，下面的代码是服务器端创建的 map 类型数据，传递到前端，使用 EL 表达式来获取。

服务器端代码：

```
Map map = new HashMap();
map.put(key1,value1);
map.put(key2,value2);
map.put(key3,value3);
```

前端视图 JSP 中的代码：

```
${map[key1]}
```

例如 map.put("name","liming")，${map["name"]} 就可以获取到 liming。

3. EL 读取 list 类型的数据

EL 可以读取 list 类型的数据，下面的代码是服务器端创建的 list 类型数据，传递到前端，使用 EL 表达式来获取。

服务器端的代码：

```
List list = new ArrayList();
list.add("aa");
list.add("bb");
list.add("cc");
```

前端视图 JSP 的代码：

```
${list[0]}
```

${list[0]} 获取到 "aa"，${list[1]} 获取到 "bb"。

4. EL 表达式的运算符号

EL 表达式可以使用各种运算符号，包括算术运算符（+、-、*、/、% 和 ()）、关系运算符（>、<、>=、<=、==、!=）、逻辑运算符（&&、||、!），这些运算符号的用法与 Java 语言相同，此外还有 empty 运算符号和 ? 运算符号。

empty 运算符号用于判断对象的取值是否为空，如果为空则结果为 true，不为空则结果为 false。其语法格式如下：

${empty 对象名}

例如：${empty sessionScope.username}，这个表达式表示判断 session 范围的 username 属性是否为空，如果为空，则表达式的值为 true，否则为 false。

${条件? 值1:值2}

例如：${PageNumber <= 1? 1:PageNumber - 1}

7.4.2　JSP 标准标签库

JSTL，即 JSP Standard Tag Library 的英文简称，是指 JSP 标准标签库，由 apache 的 jakarta

小组开发，Sun 公司推出的一种技术规范，旨在简化 JSP 代码的编写。JSTL 每个标签对应一段 Java 程序，实现了 JSP 通用的核心功能，标签的书写格式跟 HTML 标签一样。用户也可以根据需要将一些要重复使用的功能实现为自定义标签，方便重复使用。

JSTL 分为五个部分，每一个部分对应一个地址，要使用标签时，在 Eclipse 开发环境中，需要先将 JSTL 包(jstl1.2.jar)复制到项目的 lib 目录中，当前 JSTL 的版本是 1.2，支持的环境是 Servlet 2.5 和 JSP 2.1 及以上版本，然后在 JSP 文件中进行声明，格式如下：<%@ taglib uri = "地址" prefix = "前缀" %>。JSTL 五大标签库和对应的 uri、prefix 如表 7.2 所示。

表 7.2 JSTL 五大标签库和对应的 uri、prefix

标签库名称	uri	prefix
核心标签库	http://java.sun.com/jsp/jstl/core	c
国际化标签库	http://java.sun.com/jsp/jstl/fmt	fmt
sql 标签库	http://java.sun.com/jsp/jstl/sql	sql
XML 标签库	http://java.sun.com/jsp/jstl/xml	x
函数标准标签库	http://java.sun.com/jsp/jstl/functions	fn

关于 JSTL 包和相应的规范文档可以在官网上下载查看，网址是 http://tomcat.apache.org/taglibs/standard。

对于 JSTL 核心标签库，在项目开发中主要使用下面几个标签。

1. <c:if>标签

格式：

```
<c:if test = "逻辑表达式" >
    标签内容
</c:if>
```

用途：根据逻辑表达式的值进行判断，如果值为 true，则执行标签的内容。该标签通常用于控制不同条件下界面的显示，比如根据用户是否登录显示不同的界面。

例如：

```
<!--户未登录-->
<c:if test = "${empty sessionScope.username}" >
  <ul>
    <li><a href = "login.jsp">登录</a></li>
    <li><a href = "register.jsp">注册</a></li>
  </ul>
</c:if>
<!--用户已登录-->
<c:if test = "${! empty sessionScope.username}" >
```

```
  <ul>
    <li>${sessionScope.username},您好!</a></li>
    <li><a href="logout.jsp">注销</a></li>
  </ul>
</c:if>
```

也可用于控制不同用户权限的界面。

```
<!--角色是管理员的,左边显示管理功能的界面-->
<c:if test="${sessionScope.role == 1}">
    <jsp:include page="adminLeft.jsp"></jsp:include>
</c:if>
<!--角色是普通用户的,左边显示普通用户的功能-->
<c:if test="${sessionScope.role == 2}">
    <jsp:include page="left.jsp"></jsp:include>
</c:if>
```

2. <c:forEach>标签

格式：

```
<c:forEach items="集合名称" var="集合元素的变量名">
    迭代的内容
</c:forEach>
```

用途：通常用于将从数据库中查询到的记录显示在视图中,例如查询所有用户名,每一条记录封装为一个对象,然后迭代显示出每个对象的属性值,即显示数据表的一条记录的各个字段值。

```
<c:forEach items="userList" var="user">
  <tr>
    <td>${user.id}</td>
    <td>${user.username}</td>
    <td>${user.password}</td>
  </tr>
</c:forEach>
```

结合前面介绍 EL 获取 list 类型的数据,使用<c:forEach>标签迭代进行显示,代码如下。

```
<c:forEach items="list" var="l">
  <tr>
    <td>${l}</td>
  </tr>
</c:forEach>
```

第 8 章　Java EE 综合应用开发

【本章内容】
- 8.1　软件开发流程和规范
- 8.2　留言板系统设计
- 8.3　MVC 架构模式的实现

8.1　软件开发流程和规范

8.1.1　软件开发流程

软件开发流程包括软件需求分析、软件的总体结构设计、功能模块设计,以及数据库设计、编码、调试、整合和测试等过程。Java EE 课程设计的目的是让学生熟练掌握 Java EE 开发技术,能够综合地运用所学知识进行 Web 应用系统的开发,提高工程能力和培养职业素养。所以在课程设计中,应根据课程要达成的目标,由教师设计一些合适的题目,完成软件开发流程的需求分析和数据库设计,然后由学生运用所学知识进行系统架构设计、功能模块设计、编码、调试,最后进行测试并撰写相关的文档。

8.1.2　软件开发规范

在课程设计过程中,要求学生不仅要能够完成功能模块的代码编写,还要注重代码的编写质量,包括命名和注释应规范,程序结构要清晰,选用合适的架构模式。下面对命名规范和注释规范做一下介绍,架构模式的运用后面通过案例进行介绍。

1. 命名规范
①所有的命名都不要使用中文,要有一定的含义。
②项目名全部小写。
③包名全部小写,命名具有层次结构,各部分中间用圆点隔开,格式如下:
[类型名].[项目名].[包的类型名]。
包的类型有简单 Java 对象 pojo 包、数据访问层 dao 包、业务逻辑层 service 包和控制层 controller 包等。
④对于包里面的接口和实现类的命名,同一个业务逻辑,可以在接口文件名前面加上大写字母"I"或者相应的实现类名后面加上"Impl"。
⑤类名首字母大写,采用驼峰式命名,即每个单词的首字母大写,其他小写,最好每类

文件的名字后面加上相应类型的标识,如 JavaBean 类,文件名后加上"Bean",Servlet 类文件名后加上"Servlet"。如:

```
public class AverageBean{}    //求平均分的 JavaBean
```

⑥方法名首字母小写,如果名称由多个单词组成,后面每个单词的首字母都要大写。如:

```
public void queryAllBooks(){}
```

⑦变量名小写,常量名全部大写,如:

```
int count = 0;
public static final String USERNAME = "root";
```

2. 注释规范

程序中加上必要的注释,有助于增加程序的可阅读性,包括 JSP 注释和 Java 注释。JSP 注释的形式是 <!--注释信息-->,通常是在 JSP 文件或 XML 文件里进行注释。

Java 的注释是对 Java 代码进行注释,有如下三种。

①单行注释:以"//"开始,通常放在要注释的代码后面,对所在行代码进行注释。

②多行注释:以"/*"开始,以"*/"结束,放在要注释的代码上面,可以是一行或多行注释。

③文档注释:以"/**"开头,"*/"结束,对类和方法的注释采用这种格式的注释。

例如:对类的注释,设置了版权所有、文件名、作者、类的功能说明、日期、版本等信息

```
/**
* Copyright (C),2019-2022,XXX
* FileName: DateConverter.Java
* 日期格式转换
* @author LiMing
* @Date    2020-12-19
* @version  1.0
*/
```

例如:对方法的注释,应说明方法的功能、方法中的参数和返回的内容等信息。

```
/**
* 分页查询
* @param pageNum 页码
* @param pageSize   页的大小
* @return 查询到的记录
*/
```

8.2 留言板系统设计

8.2.1 功能模块划分

本章选择留言板系统作为综合案例进行介绍,留言板是网站管理员与用户交流的工具,是网站的一个基本功能。该系统分为管理员和游客,游客不用注册就可以浏览留言、发表留言;管理员登录之后可以回复留言和删除留言,留言板的功能模块划分如图 8.1 所示。

图 8.1　留言板的功能模块划分图

8.2.2 数据库设计

数据库统一命名为 messageboard,其中包含 note 和 noteadmin 两个表,note 表用于保存留言信息,而 noteadmin 用于保存管理员账号。

note 表和 noteadmin 表结构设计如图 8.2 和图 8.3。

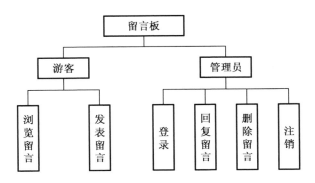

名	类型	长度	小数点	不是 null	
id	int	10	0	☑	🔑1
name	varchar	20	0	☑	
title	varchar	20	0	☑	
comment	varchar	200	0	☑	
email	varchar	30	0	☑	
url	varchar	50	0	☑	
ip	varchar	15	0	☑	
postdate	datetime	0	0	☐	
communication	varchar	50	0	☐	
isreply	int	1	0	☐	
reply	varchar	200	0	☐	
replydate	datetime	0	0	☐	

图 8.2　note 表结构设计图

图 8.3 noteadmin 表结构设计图

8.3 MVC 架构模式的实现

8.3.1 项目框架的搭建

采用 MVC 架构模式,即将系统分解为 M(model,模型) + V(view,视图) + C(controller,控制器),模型包括了表对应的持久化类 JavaBean 和实现数据库操作的 JavaBean,视图就是各个 JSP 页面,控制器由 Servlet 来实现。

①创建项目,命名为 messageboard。

②在 src 目录中创建 mvc 包,在包中创建 3 个 JavaBean 文件和 6 个 Servlet 文件,3 个 JavaBean 分别是 Note.java(note 表的持久化类)、NoteAdmin.java(noteadmin 表的持久化类)和 SqlBean.java(对数据库操作的 JavaBean);6 个 Servlet 文件分别是 queryAllNotesServlet.java(查询所有留言)、queryNoteByIdServlet.java(根据 ID 查询指定留言)、addNoteServlet.java(添加留言)、loginServlet.java(登录)、replyNoteByIdServlet.java(回复留言)、deleteNoteByIdServlet.java(删除留言)。

③在 webapp 目录中创建 JSP 文件,分别是 index.jsp(首页)、login.jsp(登录页面)、add.jsp(添加留言页面)、admin.jsp(管理页面)、reply.jsp(回复页面)和 validate.jsp(生成验证码的页面)。

8.3.2 相关的配置

①导入 JSP 标准标签库 jstl-1.2.jar 和数据库驱动包 mysql-connector-java-8.0.15.jar 到 webapp 的 WEB-INF 中的 lib 目录;

②导入 bootstrap-3.3.7-dist 到 webapp 目录下,bootstrap 版本是 3.3.7;

③导入图片到 webapp 下新创建的文件夹 image 中;

项目框架的搭建和相关的配置两个步骤完成后,项目的目录结构如图 8.4 所示。

8.3.3 模型的编写

1. note 表对应的持久化类

Note.java 代码略。

2. noteadmin 表对应的持久化类

NoteAdmin.java 代码略。

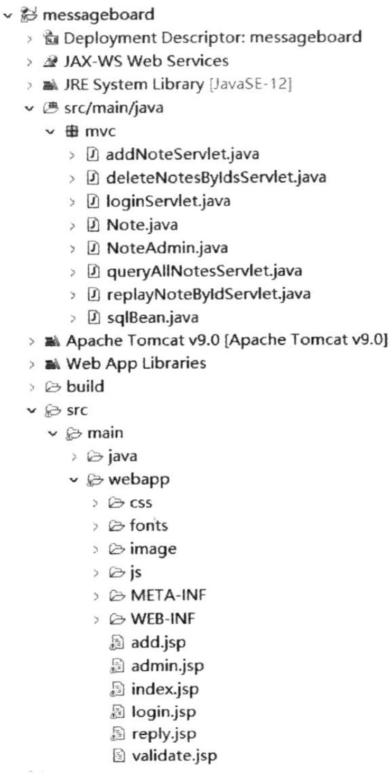

```
messageboard
  Deployment Descriptor: messageboard
  JAX-WS Web Services
  JRE System Library [JavaSE-12]
  src/main/java
    mvc
      addNoteServlet.java
      deleteNotesByIdsServlet.java
      loginServlet.java
      Note.java
      NoteAdmin.java
      queryAllNotesServlet.java
      replayNoteByIdServlet.java
      sqlBean.java
  Apache Tomcat v9.0 [Apache Tomcat v9.0]
  Web App Libraries
  build
  src
    main
      java
      webapp
        css
        fonts
        image
        js
        META-INF
        WEB-INF
        add.jsp
        admin.jsp
        index.jsp
        login.jsp
        reply.jsp
        validate.jsp
```

图8.4　项目的目录结构图

3．实现数据库操作的 JavaBean

首先通过无参构造方法创建数据库连接，即在 DirverManager.getConnection（"jdbc：mysql：//localhost：3306/messageboard？useUnicode = true&characterEncoding = UTF-8&ServerTimezone = Asia/Shanghai"，"root"，"123456"）中，通过 useUnicode = true&&characterEncoding = UTF-8 设置数据库支持中文字符，代码如下：

```java
package mvc;
import java.sql.*;
import java.util.ArrayList;
import java.util.List;
import java.util.Date;

public class SqlBean {     //SqlBean.java,实现数据库操作
    //这里可以声明 Connection 等对象
    Connection conn = null;
    PreparedStatement pstmt = null;
    ResultSet resultSet = null;
    Statement stmt = null;
    public sqlBean() {      //创建数据库连接
        try {
```

```java
        Class.forName("com.mysql.cj.jdbc.Driver");
        conn = DriverManager.getConnection("jdbc:mysql://localhost:3306/message-board? useUnicode = true&characterEncoding = UTF-8&ServerTimezone = Asia/Shanghai","root","123456");
    } catch (Exception e) {
        // TODO Auto-generated catch block
        e.printStackTrace();
    }
}
```

queryAllNotes()用于查询所有留言,通过select语句将查询到的结果保存在记录集中,然后逐条取出来,每条记录封装为一个Note对象,再添加到动态数组list中。代码如下:

```java
public List<Note> queryAllNotes() {    //查询所有留言
    List<Note> list = new ArrayList<Note>();
    String sql = "select * from note order by id desc";
    try {
        stmt = conn.createStatement();
        resultSet = stmt.executeQuery(sql);
        while (resultSet.next()) {
        Note user = new Note();
        user.setId(resultSet.getInt(1));
        user.setName(resultSet.getString(2));
        user.setTitle(resultSet.getString(3));
        user.setComment(resultSet.getString(4));
        user.setEmail(resultSet.getString(5));
        user.setUrl(resultSet.getString(6));
        user.setIp(resultSet.getString(7));
        user.setPostdate(resultSet.getDate(8));
        user.setCommunication(resultSet.getString(9));
        user.setIsreply(resultSet.getBoolean(10));
        user.setReply(resultSet.getString(11));
        user.setReplydate(resultSet.getDate(12));
        list.add(user);
        }
    } catch (SQLException e) {
        // TODO Auto-generated catch block
        e.printStackTrace();
    }
    if (conn != null) {
        try {
```

```
            conn.close();
        } catch (SQLException e) {
            //TODO Auto-generated catch block
            e.printStackTrace();
        }
    }
    return list;
}
```

queryNoteById(int id)用于查询指定的留言,方法中的参数指定查询的 id,创建预处理语句对象,将参数 id 代入,如果查询有结果,得到的记录封装为一个 Note 对象。代码如下:

```
public Note queryNoteById(int id) {    //查询指定的留言
    Note user = new Note();
    String sql = "select * from note where id = ?";
    try {
        pstmt = conn.prepareStatement(sql);
        pstmt.setInt(1, id);
        resultSet = pstmt.executeQuery();
        while (resultSet.next()) {
            user.setId(resultSet.getInt(1));
            user.setName(resultSet.getString(2));
            user.setTitle(resultSet.getString(3));
            user.setComment(resultSet.getString(4));
            user.setEmail(resultSet.getString(5));
            user.setUrl(resultSet.getString(6));
            user.setIp(resultSet.getString(7));
            user.setPostdate(resultSet.getDate(8));
            user.setCommunication(resultSet.getString(9));
            user.setIsreply(resultSet.getBoolean(10));
            user.setReply(resultSet.getString(11));
            user.setReplydate(resultSet.getDate(12));
        }
    } catch (SQLException e) {
        //TODO Auto-generated catch block
        e.printStackTrace();
    }
    return user;
}
```

addNote(Note note)用于添加留言,参数是一个 Note 对象,即将 Note 对象的各个属性值作为各个字段的值,添加一条新记录,其中 postdate 字段是添加记录的时间,通过 getTime() 获取系统的时间获得。addNote()方法返回一个 int 值,0 表示添加失败,1 表示受影响的行数为1，也就是成功地添加一条记录。代码如下：

```java
public int addNote(Note note) {    //添加留言
    int count = 0;
    Date dayDateUtil = note.getPostdate();
    Java.sql.Date dayDateSql = new Java.sql.Date(dayDateUtil.getTime());
    try {
        String sql = "insert into note(name,title,comment,email,url,ip,postdate,communication,isreply,reply,replydate) values(?,?,?,?,?,?,?,?,?,?,?)";
        pstmt = conn.prepareStatement(sql);
        pstmt.setString(1, note.getName());
        pstmt.setString(2, note.getTitle());
        pstmt.setString(3, note.getComment());
        pstmt.setString(4, note.getEmail());
        pstmt.setString(5, note.getUrl());
        pstmt.setString(6, note.getIp());
        pstmt.setDate(7, dayDateSql);
        pstmt.setString(8, note.getCommunication());
        pstmt.setInt(9, 0);
        pstmt.setString(10, null);
        pstmt.setDate(11, null);
        count = pstmt.executeUpdate();
    } catch (Exception e1) {
        // TODO Auto-generated catch block
        e1.printStackTrace();
    }
    if (conn != null && pstmt != null) {
        try {
            conn.close();
            pstmt.close();
        } catch (SQLException e) {
            // TODO Auto-generated catch block
            e.printStackTrace();
        }
    }
    return count;
}
```

replyNoteById(Note note)用于回复指定的留言，参数为Note对象，可以通过Note.getId()得到要回复记录的ID，通过Note.getReply()得到回复的内容，通过getTime()得到回复的时间，并设置isreply=1，更新指定的留言。代码如下：

```java
public int replyNoteById(Note note) {    //回复留言
    int count = 0;
    Date dayDateUtil = note.getReplydate();
    java.sql.Date dayDateSql = new java.sql.Date(dayDateUtil.getTime());
    try {
        String sql = "update note set isreply = ? , reply = ? , replydate = ? where id = ?";
        pstmt = conn.prepareStatement(sql);
        pstmt.setInt(1, 1);
        pstmt.setString(2, note.getReply());
        pstmt.setDate(3, dayDateSql);
        pstmt.setInt(4, note.getId());
        count = pstmt.executeUpdate();
    } catch (Exception e1) {
        //TODO Auto-generated catch block
        e1.printStackTrace();
    }
    if (conn != null && pstmt != null) {
        try {
            conn.close();
            pstmt.close();
        } catch (SQLException e) {
            //TODO Auto-generated catch block
            e.printStackTrace();
        }
    }
    return count;
}
```

deleteNotesByIds(int[] ids)用于删除留言,即将整型数组作为参数传入,通过 for 循环读取数组中的每一个值,然后依次删除留言,可一次删除多条留言。代码如下:

```java
public int deleteNotesByIds(int[] ids) {    //删除留言,可一次删除多条留言
    int count = 0;
    try {
        String sql = "delete from note where id = ?";
        pstmt = conn.prepareStatement(sql);
        for (int i = 0; i < ids.length; i++) {
            pstmt.setInt(1, ids[i]);
            if (pstmt.executeUpdate() != 0) {
                count++;
            }
        }
```

```java
        }
    } catch (Exception e) {
        // TODO Auto-generated catch block
        e.printStackTrace();
    }
    if (conn != null) {
        try {
            conn.close();
        } catch (SQLException e) {
            // TODO Auto-generated catch block
            e.printStackTrace();
        }
    }
    return count;
}
```

isLogin(NoteAdmin noteAdmin)用于管理员登录验证,即通过 where 子句将 admin 和 pwd 作为条件查询 noteadmin 表,如果找到跟输入的用户名和密码都相同的记录,则判断登录成功,返回 true,否则登录失败,返回 false。代码如下:

```java
public boolean isLogin(NoteAdmin noteAdmin) { //管理员登录验证
    boolean flag = false;
    String sql = "select * from noteadmin where admin=? and pwd=?";
    try {
        pstmt = conn.prepareStatement(sql);
        pstmt.setString(1, noteAdmin.getAdmin());
        pstmt.setString(2, noteAdmin.getPwd());
        resultSet = pstmt.executeQuery();
        if (resultSet.next()) {
            flag = true;
        }
    } catch (SQLException e1) {
        // TODO Auto-generated catch block
        e1.printStackTrace();
    }
    if (conn != null && pstmt != null) {
        try {
            conn.close();
            pstmt.close();
        } catch (SQLException e) {
            // TODO Auto-generated catch block
            e.printStackTrace();
```

```
            }
        }
        return flag;
}
```

8.3.4 控制器 Servlet 的编写

Servlet 程序用于控制业务流程,起到中介的作用,一方面接收客户端的请求,另一方面调用相应的业务逻辑组件处理业务,并对客户端作出响应。

1. 显示所有留言 QueryAllNotesServlet.java

调用 SqlBean.java 中的 queryAllNotes(),查询所有留言,将结果返回给 list,设置为 session 属性,然后跳转到 index.jsp 页面。代码如下:

```java
@WebServlet("/QueryAllNotesServlet")
public class QueryAllNotesServlet extends HttpServlet {  //QueryAllNotesServlet.java,用来显示所有留言
    protected void doGet(HttpServletRequest request, HttpServletResponse response)
    throws ServletException, IOException {
        //TODO Auto-generated method stub
        SqlBean sqlbean = new SqlBean();
        try {
            List<Note> list = sqlbean.queryAllNotes();
            HttpSession session = request.getSession();
            session.setAttribute("list",list);
            request.getRequestDispatcher("/index.jsp").forward(request,response);
        } catch (Exception e) {
            //TODO Auto-generated catch block
            e.printStackTrace();
        }
    }
}
```

2. 添加留言 AddNoteServlet.java

读取表单提交的留言数据,封装为 note 对象,调用 SqlBean.java 的 addNote(),如果添加成功,则设置 request 的 addSuccess 属性,然后跳转到 QueryAllNotesServlet 页面,否则设置 request 的 addFail 属性,然后跳转到 add.jsp 页面,重新添加。代码如下:

```java
@WebServlet("/AddNoteServlet")
public class AddNoteServlet extends HttpServlet {  //addNoteServlet.java,用来添加留言
    protected void doPost(HttpServletRequest request, HttpServletResponse response) throws ServletException, IOException {
        //TODO Auto-generated method stub
```

```java
        request.setCharacterEncoding("UTF-8");
        response.setCharacterEncoding("UTF-8");
        Note note = new Note();
        SqlBean sqlbean = new SqlBean();
        String name = request.getParameter("name");
        String title = request.getParameter("title");
        String comment = request.getParameter("comment");
        String url = request.getParameter("url");
        String ip = request.getParameter("ip");
        String email = request.getParameter("email");
        String communication = request.getParameter("communication");
        note.setName(name);
        note.setTitle(title);
        note.setComment(comment);
        note.setEmail(email);
        note.setUrl(url);
        note.setIp(ip);
        note.setPostdate(new Date());
        note.setCommunication(communication);
        if(sqlbean.addNote(note)!=0){//成功
            //note.setIsreply(true);
            request.setAttribute("addSuccess","发表成功!");
            request.getRequestDispatcher("/QueryAllNotesServlet").forward(request, response);
        }else{//失败
            request.setAttribute("addFail","发表失败!");
            request.getRequestDispatcher("/add.jsp").forward(request, response);
        }
    }
}
```

3. 删除留言 DeleteNotesServlet.java

获取表单提交的多值元素 id 的值,然后将各个值字符型转换为整型,调用 SqlBean.java 中的 deleteNotesByIds()删除多条记录,如果成功,则设置 request 的 delete 属性值为"删除成功",否则设置为"删除失败",然后跳转到 QueryAllNotesServlet。代码如下:

```java
@WebServlet("/DeleteNotesServlet")
public class DeleteNotesServlet extends HttpServlet {//deleteNotesServlet.java,
                                                    //用来删除留言
    protected void doPost(HttpServletRequest request, HttpServletResponse response) throws ServletException, IOException {
        // TODO Auto-generated method stub
```

```java
        String[] id = (String[])request.getParameterValues("id");
        int[] ids = new int[id.length];
        for(int i = 0; i < id.length; i++)
            ids[i] = Integer.parseInt(id[i]);
        SqlBean sqlbean = new SqlBean();
        if(sqlbean.deleteNotesByIds(ids)!=0){
            request.setAttribute("delete","删除成功");
            request.getRequestDispatcher("/QueryAllNotesServlet").forward(request, response);//跳转
            return;
        }
        else{
            request.setAttribute("delete","删除失败");
            request.getRequestDispatcher("/QueryAllNotesServlet").forward(request, response);//跳转
            return;
        }
    }
}
```

4. 回复留言 ReplayNoteByIdServlet.java

首先，获取表单元素 id 和 reply 的值，通过 SqlBean.java 中的 queryNoteById( ) 查询到指定 id 的留言，然后设置 note 的 reply、replydate 属性的值，再调用 SqlBean.java 中的 replyNoteById( )，如果成功，则设置 request 的 addSuccess 属性值，跳转到 QueryAllNotesServlet，否则设置 request 的 addFail 属性值，跳转到 reply.jsp 页面。代码如下：

```java
@WebServlet("/ReplayNoteByIdServlet")
public class ReplayNoteByIdServlet extends HttpServlet{ //回复留言
    protected void doPost(HttpServletRequest request, HttpServletResponse response) throws ServletException, IOException{
        //TODO Auto-generated method stub
        request.setCharacterEncoding("UTF-8");
        response.setCharacterEncoding("UTF-8");
        String uid = new String((String)request.getParameter("id"));
        int id = Integer.parseInt(uid);
        String reply = request.getParameter("reply");
        SqlBean sqlbean = new SqlBean();
        Note note = sqlbean.queryNoteById(id);
        note.setReply(reply);
        note.setReplydate(new Date());
        int count = sqlbean.replyNoteById(note);
        if(count!=0){//成功
```

```
            request.setAttribute("addSuccess","发表成功!");
            request.getRequestDispatcher("/QueryAllNotesServlet").forward(request,response);
        }else{ //失败
            request.setAttribute("addFail","发表失败!");
            request.getRequestDispatcher("/reply.jsp").forward(request, response);
        }
    }
}
```

5. 登录 LoginServlet.java

通过 request.getParameter() 获取表单元素 admin、pwd 和 code 的值，code 是在页面中用户输入的验证码，由后台产生的验证码保存为 session 属性 code 的值，判断两者是否一致，同时调用 SqlBean.java 中的 isLogin() 比较用户名和密码是否输入正确，以上两部分判断都为真，则表示登录验证通过，设置 request 的 loginSuccess 属性的值，并跳转到 QueryAllNotesServlet，否则，验证通不过，设置 requeste 的 loginError 属性的值，并跳转到 login.jsp 页面。代码如下：

```
@WebServlet("/LoginServlet")
public class LoginServlet extends HttpServlet{ //loginServlet.java,用来处理管理员登录
    protected void doPost(HttpServletRequest request, HttpServletResponse response) throws ServletException, IOException{
        //TODO Auto-generated method stub
        request.setCharacterEncoding("utf-8");
        response.setContentType("text/html;charset=utf-8");
        String admin = request.getParameter("admin");
        String pwd = request.getParameter("pwd");
        String code = request.getParameter("code");
        NoteAdmin user = new NoteAdmin();
        user.setAdmin(admin);
        user.setPwd(pwd);
        SqlBean sqlbean = new SqlBean();
        HttpSession session = request.getSession();
        String randStr = (String)session.getAttribute("randStr");
        if(sqlbean.isLogin(user) && code.equals(randStr)){
            request.setAttribute("loginSuccess","登录成功!");
            request.getRequestDispatcher("/QueryAllNotesServlet").forward(request, response);
        } else {
            request.setAttribute("loginError","用户名或密码或验证码错误! 请重新填写!");
```

```
            request.getRequestDispatcher("/login.jsp").forward(request,re-
sponse);
        }
    }
}
```

8.3.5 视图文件

视图文件采用 div+css 设置样式,并且采用 bootstrap 设置各个 div 的样式。

1. 显示所有留言 index.jsp

代码如下:

```
<%@ page language="java" import="mvc.sqlBean,mvc.Note,java.util.*" content-
Type="text/html;charset=utf-8" pageEncoding="utf-8"%>
    <%@ taglib uri="http://java.sun.com/jsp/jstl/core" prefix="c"%>
<!DOCTYPE html>
<html>
<head>
<meta http-equiv="Content-Type" content="text/html;charset=ISO-8859-1">
<title>留言板</title>
<link rel="stylesheet" href="/css/bootstrap.min.css">
<script src="/js/jquery.min.js"></script>
<script src="/js/bootstrap.min.js"></script>
<style type="text/css">
#head{
    background-image:url(./image/tb.jpg);
    background-size:100%;
    background-repeat:no-repeat;
    background-position:center;
    }
table{
    width:1000px;
}
#tb{
    margin:30px 70px;
}
body{
    /* background-color:#f2eada; */
```

```
        background:url(./image/lybj.jpg);
        background-size:100% 100%;
}
</style>
</head>
<body>
<div class="jumbotron text-center" style="margin-bottom:0" id="head">
  <h1>畅所欲言,畅你所言</h1>
  <p>biu biu biu~</p>
</div>
<nav class="navbar navbar-inverse">
  <div class="container-fluid">
    <div class="navbar-header">
      <button type="button" class="navbar-toggle" data-toggle="collapse" data-target="#myNavbar">
        <span class="icon-bar"></span>
        <span class="icon-bar"></span>
        <span class="icon-bar"></span>
      </button>
      <a class="navbar-brand" href="#">畅言网</a>
    </div>
    <div class="collapse navbar-collapse" id="myNavbar">
      <ul class="nav navbar-nav navbar-right">
        <li class="active"><a href="#">主页</a></li>
        <li><a href="add.jsp">发布留言</a></li>
        <li><a href="login.jsp">管理员登陆</a></li>
      </ul>
    </div>
  </div>
</nav>
<%
String successInfo = (String)request.getAttribute("addSuccess");//获取成功属性
if(successInfo != null){
%>
<script type="text/Javascript" language="Javascript">
    alert("<%=successInfo%>");                    //弹出成功信息
    window.location='index.jsp';                  //跳转到主页界面
</script>
<%
    }
```

```jsp
%>
<div class="container" id="center">
  <div class="row">
    <div class="col-sm-8">
    <table width="75%" border="1" id="tb">
        <c:forEach var="list" items="${sessionScope.list}">
        <tr>
        <td width="30%">
            <div class="text-center" style="margin:10px 10px">
                <p>留言人：${list.name}</p>
                <p>来自：${list.ip}</p>
                <p>邮箱：${list.email}</p>
                <p>主页：${list.url}</p>
                <p>${list.postdate}</p>
            </div>
        </td>
        <td width="70%">
            <p style="margin:10px 3px;margin-top:0px; margin-bottom:0px;">主题：${list.title}
            <hr style="background-color:gray; height:1px; border:none; margin-top:0px;margin-bottom:0px;">
            </p>
            <div style="margin:10px 3px;">${list.comment}</div>
            <br>
            <div style="margin:10px 3px;">
            <c:if test="${list.isreply==true}">
                <div style="margin:10px 30px;">回复：${list.name}　${list.reply}</div>
                <p style="margin:0px 30px;margin-top:0px; margin-bottom:0px;">${list.replydate}</p>
            </c:if>
            </div>
        </td>
        </tr>
        </c:forEach>
    </table>
    </div>
  </div>
</div>
</body>
</html>
```

2. 添加留言 add.jsp

代码如下：

```html
<!DOCTYPE html>
<html>
<head>
<style type="text/css">
#firstDiv{
    width:800px;
    height:400px;
    box-shadow:4px 4px 20px #909090;/* opera 或 ie9 */
}
#bnt{
    text-align: center;
}
body{
    background:url(./image/lybj.jpg);
}
#comment{
}
</style>
</head>
<body>
<%
String addFail = (String)request.getAttribute("addFail"); //获取失败属性
if(addFail != null){
%>
    <script type="text/Javascript" language="Javascript">
    alert("<%=addFail%>");                              //弹出失败信息
    </script>
<%}%>
<nav class="navbar navbar-inverse">
  <div class="container-fluid">
    <div class="navbar-header">
      <button type="button" class="navbar-toggle" data-toggle="collapse" data-target="#myNavbar">
        <span class="icon-bar"></span>
        <span class="icon-bar"></span>
        <span class="icon-bar"></span>
      </button>
```

```html
        <a class="navbar-brand" href="#">畅言网</a>
      </div>
      <div class="collapse navbar-collapse" id="myNavbar">
        <ul class="nav navbar-nav navbar-right">
          <li><a href="index.jsp">主页</a></li>
          <li class="active"><a href="#">发布留言</a></li>
        </ul>
      </div>
    </div>
  </nav>
  <br><br>
  <div class="container" id="firstDiv" background="white">
  <h2 style="text-align:center">请输入您的留言:</h2>
    <div class="row">
    <br><br>
    <form action="AddNoteServlet" method="post">
      <table border="1" align="center" height="100" width="400" class="table table-bordered">
          <tr>
            <td style="text-align:right" width="40%">您的名字</td>
            <td width="60%"><input type="text" name="name" placeholder="*必须填写" id="name"></td>
          </tr>
          <tr>
            <td style="text-align:right">留言主题:</td>
            <td><input type="text" name="title" placeholder="*必须填写" id="title"></td>
          </tr>
          <tr>
            <td style="text-align:right">您的邮箱:</td>
            <td><input type="text" name="email" id="text"></td>
          </tr>
          <tr>
            <td style="text-align:right">其他联系方式:</td>
            <td><input type="text" name="communication" id="communication"></td>
          </tr>
          <tr>
            <td style="text-align:right">留言内容:</td>
            <td><textarea row="3" cols="25" name="comment" placeholder="*100字以内" id="comment"></textarea></td>
```

```
            </tr>
            <tr class="sr-only">
                <td height="93">url地址:</td>
                <td colspan="2"><input name="url" type="hidden" value="<%=url%>"/></td>
            </tr>
            <tr class="sr-only">
                <td height="93">Ip地址:</td>
                <td colspan="2"><input name="ip" type="hidden" value="<%=request.getRemoteAddr()%>"/></td>
            </tr>
        </table>
        <div id="bnt">
            <input type="submit" value="提交留言" class="btn btn-primary" style="margin-right:20px">
            <input type="reset" value="重新编写" class="btn btn-primary" id="resetBtn">
        </div>
    </form>
  </div>
 </div>
</body>
</html>
```

3. 登录页面 login.jsp

代码如下:

```
<%@ page language="java" contentType="text/html;charset=utf-8"
    pageEncoding="utf-8"%>
<!DOCTYPE html>
<html>
<head>
<style type="text/css">
#firstDiv{
    width:800px;
    height:400px;
    margin:70px 350px;
}
#bg{
    width:600px;
    height:300px;
    background-color:white;
```

```css
        box-shadow:4px 4px 20px #909090;/*opera 或 ie9*/
}
#bnt{
    text-align: center;
}
body{
    background:url(./image/dltp.jpg);
    background-size: cover;
}
</style>
</head>
<body>
<script type="text/Javascript">
function refresh(){
    login.imgValidate.src = "validate.jsp? id=" + "Math.random()";
}
</script>
    <%
String errorInfo = (String)request.getAttribute("loginError");    //获取错误属性
if(errorInfo != null){
%>
    <script type="text/Javascript" language="Javascript">
alert("<%=errorInfo%>");                                          //弹出错误信息
window.location=login.jsp;                                        //跳转到登录界面
</script>
    <%
}
%>
<nav class="navbar navbar-inverse">
  <div class="container-fluid">
    <div class="navbar-header">
      <button type="button" class="navbar-toggle" data-toggle="collapse" data-target="#myNavbar">
        <span class="icon-bar"></span>
        <span class="icon-bar"></span>
        <span class="icon-bar"></span>
      </button>
      <a class="navbar-brand" href="#">畅言网</a>
    </div>
    <div class="collapse navbar-collapse" id="myNavbar">
```

```html
    <ul class="nav navbar-nav navbar-right">
      <li><a href="index.jsp">返回主页</a></li>
    </ul>
   </div>
  </div>
 </nav>
<br>
<br><br>
<div class="container" id="firstDiv">
<div id="bg">
<br>
<h2 style="text-align:center">管理员登陆</h2>
  <div class="row">
  <br>
  <form action="LoginServlet" method="post">
    <table border="1" align="center" height="100" width="400" class="table-bordered" id="tb">
      <tr>
        <td style="text-align:right" width="200">用户名:</td>
        <td><input type="text" name="admin" id="admin"></td>
      </tr>
      <tr>
        <td style="text-align:right">密码:</td>
        <td><input type="password" name="pwd" id="pwd"></td>
      </tr>
      <tr>
        <td style="text-align:right">验证码:</td>
        <td><input type="text" name="code" id="code" size="10"><img name="imgValidate"
         src="validate.jsp" onclick="refresh()"></td>
      </tr>
    </table>
    <br><br>
      <div id="bnt">
      <input type="submit" value="登陆" class="btn btn-primary" style="margin-right:20px">
        <input type="reset" value="重置" class="btn btn-primary" id="resetBtn">
      </div>
    </form>
```

```
        </div>
      </div>
   </div>
</body>
</html>
```

4. 管理页面 admin.jsp

代码如下:

```jsp
<%@ page language="java" contentType="text/html;charset=utf-8"
    pageEncoding="utf-8"%>
<%@ taglib uri="http://java.sun.com/jsp/jstl/core" prefix="c"%>
<!DOCTYPE html>
<html>
<head>
<style type="text/css">
.a{
    color:gray;
    cursor:pointer;
    pointer-events: none;
}
#div2{
    margin:15px 70px;

}
#div3{
    margin:4px 3px;
}
table{
    background-color:
}
body{
    background:url(./image/adbj.jpeg);
    background-size: cover;
    /* height:800px; */
}
</style>
</head>
<body>
<%
    String successInfo = (String) request.getAttribute("loginSuccess"); //获取成功属性
    if (successInfo != null) {
```

```
    %>
    <script type="text/Javascript" language="Javascript">
alert("<%=successInfo%>");                    //弹出成功信息
window.location='admin.jsp';                  //跳转到管理界面
    </script>
    <%
    }
    %>
    <%
    String Info = (String)request.getAttribute("delete");  //获取成功属性
    if(Info != null){
    %>
    <script type="text/Javascript" language="Javascript">
alert("<%=Info%>");    //弹出成功信息
window.location='admin.jsp';                              //跳转到管理界面
    </script>
    <%
    }
    %>
<!--删除框提示-->
<script language="Javascript">
function sumbit_sure(){
    var checkedNum = $("input[name='id']:checked").length;
    if(checkedNum==0){
        alert("请至少选择一项!");
        return false;
    }
    var gnl=confirm("确定要删除?");
    if(gnl==true){ return true;
    }
    else{
        return false;
    }
}
</script>
<nav class="navbar navbar-inverse">
  <div class="container-fluid">
    <div class="navbar-header">
      <button type="button" class="navbar-toggle" data-toggle="collapse" data-target="#myNavbar">
```

```html
        <span class="icon-bar"></span>
        <span class="icon-bar"></span>
        <span class="icon-bar"></span>
      </button>
      <a class="navbar-brand" href="#">畅言网</a>
    </div>
    <div class="collapse navbar-collapse" id="myNavbar">
      <ul class="nav navbar-nav navbar-right">
        <li><a href="index.jsp">主页</a></li>
        <li class="active"><a href="#">管理员界面</a></li>
        <li><a href="login.jsp">退出</a></li>
      </ul>
    </div>
  </div>
</nav>
<div class="text-center" id="div2">
<form action="DeleteNotesServlet" method="post" onsubmit="return sumbit_sure()">

  <table width="100%" border="1">
    <tr>
        <td height="45">选项</td>
        <td height="45">留言ID号</td>
        <td height="45">留言者姓名</td>
        <td height="45">留言的主题</td>
        <td height="45">留言的内容</td>
        <td height="45">留言提交的日期</td>
        <td height="45">留言者的email地址</td>
        <td height="45">留言者的url地址</td>
        <td height="45">留言者的IP地址</td>
        <td height="45">留言者的其他联系方式</td>
        <td height="45">回复的内容</td>
        <td height="45">回复的日期</td>
        <td height="45">是否回复</td>
    </tr>
    <c:forEach items="${sessionScope.list}" var="note" varStatus="status">
        <tr>
            <td height="45"><input type="checkbox"
                name="id" value="${note.id}" /></td>
            <td height="45">${note.id}</td>
            <td height="45">${note.name}</td>
```

```jsp
                <td height="45">${note.title}</td>
                <td height="45">${note.comment}</td>
                <td height="45">${note.postdate}</td>
                <td height="45">${note.email}</td>
                <td height="45">${note.url}</td>
                <td height="45">${note.ip}</td>
                <td height="45">${note.communication}</td>
                <td height="45">${note.reply}</td>
                <td height="45">${note.replydate}</td>
                <%-- <td height="45" bgcolor="#99cc66">${note.isreply}</td> --%>
                <td height="45">
                  <c:if test="${note.isreply==true}">已回复</c:if>
                  <c:if test="${note.isreply==false}"><a href="reply.jsp?id=${note.id}&name=${note.name}">未回复</a></c:if>
                </td>
            </tr>
        </c:forEach>
    </table>
    <div style="text-align:left" id="div3">
        <input type="submit" value="删除" class="btn btn-primary" style="margin-right:20px">
    </div>
    </form>
</div>
</body>
</html>
```

5. 回复留言 reply.jsp

代码如下:

```jsp
<%@page language="java" contentType="text/html;charset=utf-8"
    pageEncoding="utf-8"%>
<!DOCTYPE html>
<html>
<head>
<style type="text/css">
#firstDiv{
    width:800px;
    height:400px;
}
#bg{
    width:600px;
```

```css
    height:370px;
    box-shadow:4px 4px 20px #909090;/* opera 或 ie9 */
    margin:20px 60px;
}
#bnt{
    text-align: center;
}
body{
    background:url(./image/lybj.jpg);
}
textarea{
    height:150px;
        width:250px;
}
.textarea{
    height:150px;
        width:300px;
}
#tb{
    width:400px;
}
</style>
</head>
<body>
<nav class = "navbar navbar-inverse" >
  <div class = "container-fluid" >
    <div class = "navbar-header" >
      <button type = "button" class = "navbar-toggle" data-toggle = "collapse" data-target = "#myNavbar" >
        <span class = "icon-bar" ></span>
        <span class = "icon-bar" ></span>
        <span class = "icon-bar" ></span>
      </button>
      <a class = "navbar-brand " href = "#" >畅言网</a>
    </div>
    <div class = "collapse navbar-collapse" id = "myNavbar" >
      <ul class = "nav navbar-nav navbar-right" >
        <li><a href = "index.jsp" >主页</a></li>
        <li class = "active" ><a href = "#" >管理员界面</a></li>
        <li><a href = "/QueryAllNoteServlet" >退出</a></li>
```

```html
        </ul>
      </div>
   </div>
</nav>
<br>
<div class="container" id="firstDiv">
<div id="bg"><br>
<h2 style="text-align:center">管理员回复</h2>
   <div class="row">
   <form action="ReplayNoteServlet" method="post">
   <table border="1" align="center" height="100" width="200" class="table table-bordered" id="tb">
   <tr>
       <td colspan="2"><input type="hidden" name="id"
          value="<%=id%>"/></td>
   </tr>
   <tr>
       <td width="28%" height="36">回复:</td>
       <td colspan="2"><input type="text" name="name"
           value="<%=name%>" readonly="readonly" style="border:0;outline:0;"/></td>
   </tr>
   <tr>
       <td height="93">回复留言:</td>
       <td colspan="2"><textarea name="reply" row="8" cols="30" style="border:0;outline:0;"></textarea></td>
   </tr>
   </table>
       <div id="bnt">
        <button type="submit" class="btn btn-primary" style="margin-right:20px">回复</button>
        <button type="reset" class="btn btn-primary" id="resetBtn">重置</button>
       </div>
     </form>
    </div>
    </div>
  </div>
</body>
</html>
```

8.3.6 项目部署和测试

1. 项目部署

将 messageboard 项目部署到服务器后,在浏览器地址栏输入 http:// localhost:8080/messageboard,运行 index.jsp 页面,结果如图 8.5 所示。

图 8.5 项目首页

2. 发布留言

点击首页右上角的"发布留言",链接到 addNote.jsp 页面,如图 8.6 所示。

图 8.6 发布留言页面

如图 8.7 所示,填写您的名字、留言主题、您的邮箱、其他联系方式和留言内容,点击"提交留言"按钮,显示"发布成功",如图 8.8 所示,点击"确定"按钮,跳转到 index.jsp。

图 8.7 发布留言页面

图 8.8 发布成功的提示信息

如图 8.9 所示,在主页上,可以看到增加了新发布的留言。

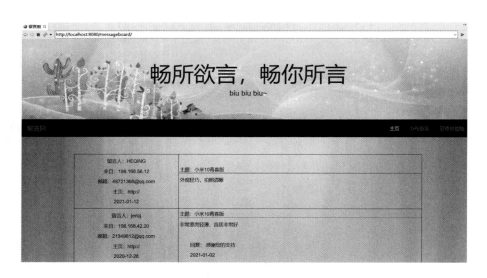

图 8.9 主页上的留言

3. 管理留言

点击主页右上角的"管理员登陆",链接到管理员登录页面 login.jsp,如图 8.10 所示,填写管理员用户名和密码,并填写验证码。

第8章　Java EE综合应用开发

图 8.10　管理员登录页面

登录验证成功,进入管理页面 admin.jsp,如图 8.11 所示,可以看到刚添加的留言在最上面,并且在"是否回复"这一栏中显示"未回复"。

选项	留言ID号	留言者姓名	留言的主题	留言的内容	留言提交的日期	留言者的email地址	留言者的url地址	留言者的IP地址	留言者的其他联系方式	回复的内容	回复的日期	是否回复
☐	6	HEQING	小米10青春版	外观轻巧,拍照清晰	2021-01-12	48721368@qq.com	http://	198.168.56.12	HEQING@tom.com			未回复
☐	5	jericj	小米10青春版	非常漂亮轻薄,音质非常好	2020-12-28	21349812@qq.com	http://	198.168.42.20	21349812@qq.com	感谢您的支持	2021-01-02	已回复
☐	4	山不高水长流	小米10青春版	大小合适,白色靓丽,拍照清晰好用	2020-12-26	186534886@qq.com	http://	198.168.65.14	xmefngkj@qq.com	您的好评是我们前进的动力	2021-01-02	已回复
☐	3	dxsc9	小米10青春版	运行顺畅,指纹识别速度快,面部解锁也很快	2020-12-26	156486454@qq.com	http://	198.168.50.5	164864536	感谢您的支持	2021-01-01	已回复

删除

图 8.11　管理页面

点击最上面留言的"未回复",进入到回复页面 reply.jsp。如图 8.12 所示,填写回复的内容,点击回复按钮。

图 8.12　管理员回复页面

然后跳转到管理页面,如图 8.13 所示,可以看到最上面的那条留言在"是否回复"栏中已显示"已回复"。

在管理页面中，勾选要删除的记录，然后单击"删除"按钮，可以一次删除多条记录，如图 8.13 所示。

图 8.13　选择要删除的记录

删除前弹出确认对话框，如图 8.14 所示，提示"确定要删除？"，如果确定，则单击"确定"按钮。

图 8.14　删除确认对话框

删除完成后回到管理页面，如图 8.15 所示，可以看到选中的记录已被删除。

图 8.15　管理页面

第 9 章　Spring 框架和 SpringMVC 框架

【本章内容】

- 9.1　Spring 框架初识
- 9.2　SpringMVC 框架概述
- 9.3　SpringMVC 框架详细介绍
- 9.4　SpringMVC 数据校验
- 9.5　SpringMVC 文件上传

9.1　Spring 框架初识

9.1.1　Spring 框架简介

Spring 框架是由 Rob Johnson 组织和开发的一种轻量级的 Java 开源框架,致力于解决企业应用开发的复杂性。Spring 框架采用分层架构,集成 Java EE 开发各层的优秀框架,以 IoC(控制反转)和 AOP(面向切面编程)为核心,使用 IoC 容器负责创建和管理各层组件,从而降低各组件之间的耦合度;通过使用 AOP,将分散在系统各业务逻辑中的辅助功能模块集中起来,同样也降低了系统的耦合度,使系统更便于管理和维护。

9.1.2　Spring 框架的体系结构

Spring 框架采用分层架构,本书采用 Spring 5.3.6 版本,其体系结构如图 9.1 所示,由下往上分为 Test、Core Container、AOP、Aspects、Instrumentation、Messaging、Data Access/Integration 和 Web 五大部分,其中黑色的部分是常用的模块。

1. Core Container 部分

即核心容器,是其他模块的基础,包括 Beans、Core、Context 和 SpEL 共四个模块。

①Beans 模块:Spring 框架的基础模块,采用工厂设计模式,由 BeanFactory 接口的实现类实现对各种 Bean 的管理,是基础的 IoC 容器。

②Core 模块:Spring 框架的核心模块,提供了 IoC 和 DI(依赖注入)的功能。

③Context 模块:建立在 Beans 模块和 Core 模块之上,提供 ApplicationContext 接口,扩展 BeanFactory 接口的功能,添加了对 Bean 的生命周期控制、框架事件体系及资源加载透明化等功能,还提供了对国际化和 JNDI、集成 EJB、电子邮件等企业级服务的支持。

④SpEL 模块:Spring 框架的表达式语言,可以查询、管理运行中的对象,方便调用对象

方法、操作数组、集合等,语法类似于 JSP 的 EL。

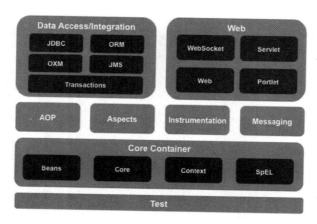

图 9.1　Spring 5.3.6 框架的体系结构

2. Data Access/ Integration 部分

即数据访问和集成,是数据访问层的模块,可以同多个流行的 ORM 框架集成在一起,包括 JDBC 模块、ORM 模块、OXM 模块、JMS 模块和 Transactions 模块。

①JDBC 模块:提供一个 JDBC 抽象层,减少对数据库操作的编码,主要是提供 JDBC 模板方式、关系数据库对象化方式、SimpleJDBC 方式、事务管理来简化 JDBC 编程。

②ORM 模块:对流行的对象关系映射框架如 Java 官方提供的 JPA、Hibernate、Mybatis 提供集成层的支持。

③OXM 模块:提供一个支持对象/ XML 映射的抽象层实现,如 JAXB、Castor、XML-Beans、JiBX 和 XStream。

④JMS 模块:Spring 框架提供了一个 JMS 集成框架,简化了 JMS API 的使用,org.springframework.jms.core 提供了使用 JMS 的核心功能,包含 JmsTemplate 类。

⑤Transactions 模块:事务模块,支持对实现特殊接口,以及所有 POJO 类的编程和声明式的事务管理。

3. Web 部分

包括 WebSocket 模块、Servlet 模块、Web 模块和 Portlet 模块,其中 WebSocket 模块和 Portlet 模块是 Spring 4.1 新增的模块,在一般的 Web 应用项目中不常用。

①Web 模块:提供基本的 Web 应用支持,如文件上传、使用 Servlet 监听器初始化 IoC 容器及 Web 应用上下文。

②Servlet 模块:提供一个 Spring 框架的 MVC 实现模块,即 SpringMVC 框架。

③WebSocket 模块:提供了对 WebSocket 协议、SockJS 协议及 STOMP 协议的支持。WebSocket 协议可实现客户端和服务器端的全双工通信,适合用于低延迟、高频消息通信的场景。SockJS 协议是 WebSocket 协议的降级方案,可以在不支持 WebSocket 协议的场景中使用。STOMP(Simple/ Streaming Text Orientated Messaging Protocol,简单流文本定向消息协

议)提供了一个可互操作的连接格式,允许 STOMP 客户端与任意 STOMP 消息代理(Broker)进行交互,项目中实时消息通知的功能用到了 STOMP 协议。

④Portlet 模块:提供 Portlet 环境下的 MVC 实现。Portlets 是一种 Web 组件,就像 Servlets 是专为将合成页面里的内容聚集在一起而设计的。通常请求一个 Portal 页面会引发多个 Portlets 被调用。每个 Portlet 都会生成标记段,并与别的 Portlets 生成的标记段组合在一起嵌入到 Portal 页面的标记内。

4. 其他模块

①AOP 模块:Spring 框架的一个核心模块,是 AOP 主要的实现模块。

②Aspects 模块:提供和 AspectJ 框架的整合,AspectJ 是一个功能强大且成熟的 AOP 框架。

③Test 模块:提供了对 Junit 和 TestNG 进行单元测试和集成测试的支持。

9.1.3 Spring 框架包的下载及介绍

在 https://repo.spring.io/simple/libs-release-local/org/springframework/spring/5.3.6/ 下载 Spring 5.3.6 框架的压缩包,打开的页面如图 9.2 所示,选择第一项下载。

图 9.2 Spring 5.3.6 下载页面

下载压缩包后进行解压,Spring 框架的目录结构如图 9.3 所示。

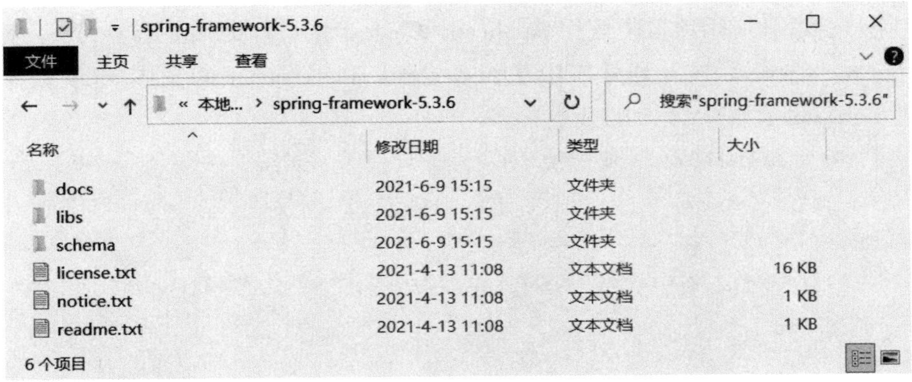

图 9.3 Spring 框架的目录结构

①docs 目录包含 Spring 框架的 API 文档和开发规范。

②在 libs 目录中，每一个 Spring 模块都包括三种类型的 JAR 包，如图 9.4 所示，例如对于 AOP 模块，包括 spring-aop-5.3.6.jar（class 文件的 JAR 包）、spring-aop-5.3.6-javadoc.jar（API 文档的压缩包）和 spring-aop-5.3.6-sourcesS.jar（源文件的压缩包）。

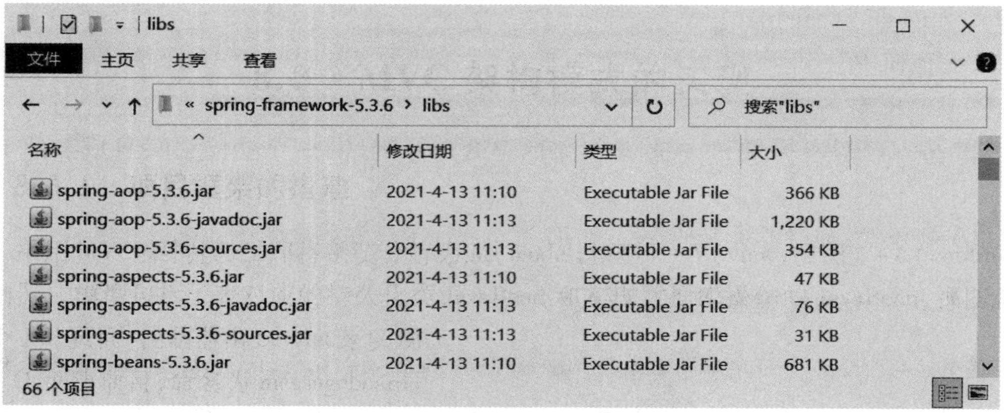

图 9.4 libs 目录结构

需要注意的是，在使用 Spring 框架开发时，除了 Spring 框架自带的 jar 包以外，还需要一个第三方 jar 包 commons.logging，用于处理日志信息。

③schema 目录包含开发所需要的 schema 文件，这些 schema 文件是整个 Spring 框架的各种配置文件，即 XML 类型的约束文件。

9.1.4 Spring 框架的核心容器

Spring 框架提供了两种核心容器实现 IoC 功能：BeanFactory 接口和 ApplicationContext 接口。BeanFactory 接口是基础的 IoC 容器，可实现创建 Bean、初始化 Bean、使用 Bean 和销毁 Bean 的功能。ApplicationContext 接口扩展了 BeanFactory 接口的功能，提供了对国际化、系统生命周期事件的支持，还提供很多对 Java 企业级服务的支持，如访问 JNDI、集成 EJB、电子邮件等。

一般在 Java EE 应用项目中使用 ApplicationContext 接口作为核心容器，Web 服务器实例化 ApplicationContext 接口，通常使用基于 ContextLoaderListener 的方式，也就是在项目的配置文件 web.xml 文件中，添加以下代码。

```
<!--配置监听器,用于加载 spring 配置文件-->
<context-param>
    <param-name>contextConfigLocation</param-name>
    <param-value>classpath:applicationContext.xml</param-value>
</context-param>
<listener>
```

```
    <listener-class>
        Org.springframework.web.context.ContextLoaderListener
    </listener-class>
</listener>
```

通过设置上下文参数的形式指明了 Spring 框架配置文件 applicationContext.xml 的位置是在类路径的根目录下,然后指定使用 ContextLoaderListener 监听器的方式来加载配置文件并实例化 ApplicationContext 接口,至于配置文件 applicationContext.xml 的内容,在 SSM 整合的章节中再做详细的介绍。

9.1.5 控制反转

IoC(Inversion of Control,控制反转)指的是应用程序中组件间的调用不再是在程序内部编码实现,而是交给外部的 Spring 容器来管理,由 Spring 容器在程序运行时根据需要来创建和销毁对象,并注入到调用它的组件。该方式也称为依赖注入(DI,Dependency Injection)。

我们来看一下前面介绍的分层架设的系统,由下至上包括数据库层、数据访问层、业务逻辑层、表现层,表现层包括视图和控制器。数据访问层、业务逻辑层和控制器都采用先定义接口,再编写实现类的方式,上下层之间的调用是在程序代码中调用的。例如在实现留言管理的业务逻辑层实现类 NoteService 的代码中,通过 new 的方式创建数据访问层 NoteDAO 类的实例化对象来进行调用,这样就造成了上下层组件之间的耦合过于紧密,代码如下。

```
public class NoteService implements INoteService {
    NoteDAO noteDAO = new NoteDAO();    //创建 NoteDAO 的实例化对象
    @Override
    //查询所有留言,返回给 admin.jsp 页面
    public List<Note> QueryAllNote() throws Exception {
        List<Note> list = noteDao.QueryAllNotes();//调用 noteDAO 的 queryAllNotes 方法
```

在 Spring 框架中,将各层组件的创建等管理工作交给容器来处理,也就是把控制权交给容器,这就是控制反转。因此可在 Spring 框架配置文件 applicationContext.xml 中对各种组件进行配置,然后由 IoC 容器进行管理,代码如下。

```
<?xml version="1.0" encoding="UTF-8"?>
<beans xmlns="http://www.springframework.org/schema/beans"
    xmlns:xsi="http://www.w3.org/2001/XMLSchema-instance"
    xsi:schemaLocation="http://www.springframework.org/schema/beans
    http://www.springframework.org/schema/beans/spring-beans-4.3.xsd">
    <!--添加各种 Bean 的配置信息,id 设置组件的实例化名称,class 指明组件的类-->
    <bean id="noteDAO" class="dao.NoteDAO">
    </bean>
    <bean id="noteService" class="service.NoteService">
        <property name="noteDAO" ref="noteDAO"/>
    </bean>
```

......
</beans>

在这个配置文件中,配置了两个组件 noteDAO 和 noteService,其中 noteService 组件要引用 noteDAO 组件,这表示 noteService 组件要调用 noteDAO 组件。

9.2 SpringMVC 框架概述

9.2.1 SpringMVC 概述

SpringMVC 框架是 Spring 框架中用于 Web 应用开发的一个模块,是实现 MVC 模式的框架。前面章节已介绍了 MVC 模式,一个系统如果采用 MVC 模式,那么就被分为三个部分,即 M(模型)、V(视图)、C(控制器)。控制器通常是由 Servlet 程序来充当,在 Servlet 程序中由开发者编写代码接收前端参数和编写代码实现视图跳转。采用 SpringMVC 框架所实现的 MVC 模式,其在参数传递和视图跳转等方面在框架中都有固定的模式,并且集成了很多 Web 应用,如文件上传、数据校验等,大大提高了开发效率,成为主流的 MVC 框架。在 Java EE 技术中,还有其他的 MVC 框架,如官方提供的 JSF(Java Server Faces),是一种用于构建基于 Java 的 Web 应用程序服务器端用户接口组件的框架,也是一个 MVC 实现框架。在主流的 SSH 框架(Spring+Struts2+Hibernate)中,其中 Struts2 框架也是一个实现 MVC 模式的框架。SpringMVC 框架作为 Spring 框架的一个部分,相比 Struts2,在使用和性能方面更加优异。

SpringMVC 框架采用前端控制器的设计模式,前端控制器作为核心的控制器,是整个框架的核心,它负责拦截客户端的请求,然后将请求映射到用户编写的业务控制器,由控制器调用业务逻辑,再调用数据访问层的方法,执行完成后将数据返回给视图,视图渲染后返回给浏览器。

9.2.2 第一个 SpringMVC 应用。

【例 9.1】 实现用户登录功能

整个项目的目录结构如图 9.5 所示。

①创建项目,选择 Dynamic Web Project 类型,导入 8 个包,其中 spring-aop-5.3.6.jar、spring-beans-5.3.6.jar、spring-context-5.3.6.jar、spring-core-5.3.6.jar、spring-expression-5.3.6.jar 是 Spring 框架的核心包,commons-logging-1.2.jar 是它们的依赖包,spring-web-5.3.6.jar 和 spring-webmvc-5.3.6.jar 是 SpringMVC 框架的包。

②在 web.xml 中配置 SpringMVC 框架的前端控制器 DispatcherServlet,代码如下。这个 Servlet 命名为"springmvc",这时 SpringMVC 框架配置文件的命名就默认为"springmvc-servlet.xml",也就是在名称后面加上"-servlet",默认的保存位置是在 WEB-INF 目录中。

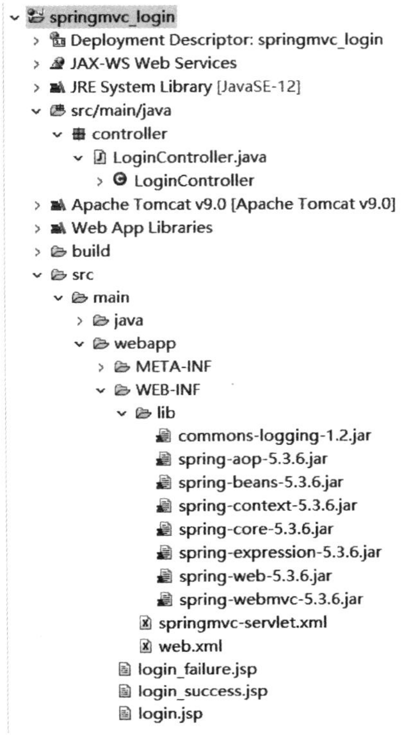

```
springmvc_login
    Deployment Descriptor: springmvc_login
    JAX-WS Web Services
    JRE System Library [JavaSE-12]
    src/main/java
        controller
            LoginController.java
                LoginController
    Apache Tomcat v9.0 [Apache Tomcat v9.0]
    Web App Libraries
    build
    src
        main
            java
            webapp
                META-INF
                WEB-INF
                    lib
                        commons-logging-1.2.jar
                        spring-aop-5.3.6.jar
                        spring-beans-5.3.6.jar
                        spring-context-5.3.6.jar
                        spring-core-5.3.6.jar
                        spring-expression-5.3.6.jar
                        spring-web-5.3.6.jar
                        spring-webmvc-5.3.6.jar
                    springmvc-servlet.xml
                    web.xml
                login_failure.jsp
                login_success.jsp
                login.jsp
```

图 9.5 项目的目录结构

```xml
<servlet>
    <servlet-name>springmvc</servlet-name>
    <servlet-class>org.springframework.web.servlet.DispatcherServlet</servlet-class>
</servlet>
<servlet-mapping>
    <servlet-name>springmvc</servlet-name>
    <url-pattern>/</url-pattern>
</servlet-mapping>
```

③在 WEB-INF 目录中，创建 SpringMVC 框架配置文件 springmvc-servlet.xml，代码如下。

```xml
<?xml version="1.0" encoding="UTF-8"?>
<beans xmlns="http://www.springframework.org/schema/beans"
    xmlns:xsi="http://www.w3.org/2001/XMLSchema-instance"
    xmlns:context="http://www.springframework.org/schema/context"
    xmlns:mvc="http://www.springframework.org/schema/mvc"
    xsi:schemaLocation="http://www.springframework.org/schema/beans
    http://www.springframework.org/schema/beans/spring-beans-4.3.xsd
    http://www.springframework.org/schema/context
    http://www.springframework.org/schema/context/spring-context-4.3.xsd
```

```
        http://www.springframework.org/schema/mvc
        http://www.springframework.org/schema/mvc/spring-mvc-4.3.xsd">
        <!--支持注解-->
        <mvc:annotation-driven/>
        <!--注解自动扫描,指定扫描controller包-->
        <context:component-scan base-package="controller"/>
        <!--配置视图解析器-->
        <bean id="viewResolver" class="org.springframework.web.servlet.view.InternalResourceViewResolver">
            <!--设置前缀-->
            <property name="prefix" value="/"/>
            <!--设置后缀-->
            <property name="suffix" value=".jsp"/>
        </bean>
</beans>
```

④在 src 目录中创建控制器的包 controller,在包中创建实现登录的控制器类,命名为 LoginController,代码如下。

```
package controller;
import javax.servlet.http.HttpSession;
import org.springframework.stereotype.Controller;
import org.springframework.web.bind.annotation.RequestMapping;
@Controller
public class LoginController {
    @RequestMapping(value="/login.do",method="post")
    public String login(String username,String password,HttpSession session){
        if(username.equals("admin")&&password.equals("123456")){
            session.setAttribute("username",username);
            return "login_success";
        }else{
            return "login_failure";
        }
    }
}
```

在控制器类中,通过@Controller 注解控制器,表示 LoginController 类是一个控制器类,@RequestMapping 注解请求映射,将 login()方法映射为一个 URL,value 参数设置请求映射 URL 为"login.do",method 参数设置对应的请求方式必须是"post"方式,也就是说当请求地址是"login.do",请求方式是"post"方式时,就会映射到"login()"方法。

⑤创建三个视图文件。

a. login.jsp 文件

```
<body>
   <form name="form1" method="post" action="login.do">
      username:<input type="text" name="username"><br>
      password:<input type="password" name="password"><br>
      <input type="submit" value="登录">
   </form>
</body>
```

b. login_success.jsp 文件

```
<body>
   你好,${sessionScope.username},login success!
</body>
```

在页面中,使用 EL 方式读取 session 域变量 username 属性的值。

c. login_failure.jsp 文件

```
<body>
   login failure!
</body>
```

⑥运行项目。

将项目部署到服务器并运行,在浏览器地址栏中输入 http://localhost:8080/springmvc_login/login.jsp,如图 9.6 所示,在登录表单中填入用户名和密码,点击"登录"按钮,则可以看到浏览器地址栏变为 http://localhost:8080/springmvc_login/login.do,即表单的用户名和密码做为参数提交给 login.do。login.do 对填写的用户名和密码进行判断,如果填写正确,则转到 login_success.jsp,在页面中显示登录的用户名等信息,如图 9.7 所示,否则转到login_failure.jsp。

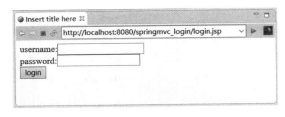

图 9.6 登录页面

图 9.7 login.do 处理结果

9.2.3 SpringMVC 的工作流程

通过上面的例子,我们已经掌握了 SpringMVC 框架应用的开发过程,接下来分析 SpringMVC 框架的工作流程,如图 9.8 所示。

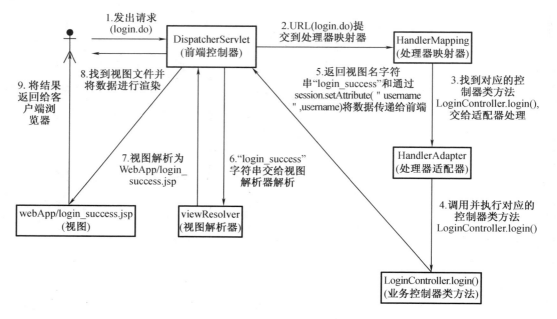

图 9.8 SpringMVC 工作流程图

①当用户发出 login.jsp 请求时,由于在 SpringMVC 框架的配置文件 springmvc-servlet.xml 中前端控制器 DisPatcherServlet 的访问地址设置为" <url-pattern>/</url-pattern>",则会匹配所有的请求,拦截静态资源和控制器类,但不会拦截 Servlet 文件和 JSP 文件,所以 login.jsp 不会被 DispatcherServlet 拦截,可以直接访问到。如果设置为" <url-pattern>/ * </url-pattern>",则会匹配所有的请求,拦截静态资源、JSP 文件和控制器类,但不拦截 Servlet 文件。当用户填写好用户名和密码,单击"登录"按钮提交给 login.do 处理时,login.do 会被 DipatcherServlet 拦截,也可以设置为 <url-pattern> *.do </url-pattern>,表示对所有.do 结尾的资源的请求都会被 DispatcherServlet 拦截。

②login.do 被拦截后交给处理器映射器 HandlerMapping 处理,HandlerMapping 有多种处理方式。当 SpringMVC 框架的配置文件设置注解支持时,同时在控制器类文件中添加了@Controller注解控制器类,则使用 HandlerMapping 接口的 RequestMappingHandler 实现类来处理。在类方法的上面添加了@RequestMapping(value = "login.do"),注解请求映射名,则 HandlerMapping 会根据请求 URL 找到对应的控制器类方法。在 springmvc_login 例子中,请求的 URL 是"login.do",所以可以通过 HandlerMapping 找到 LoginController.login()方法。

③找到对应的控制器类方法 LoginController.login(),交给处理器适配器 HandlerAdapter 处理。

④HandlerAdapter 调用并执行控制器类方法 LoginController.login()。

⑤执行完控制器类方法后,HandlerAdapter 将 ModelAndView 对象返回给前端控制器 DispatcherServlet,其中包含了模型和视图名。在 springmvc_login 例子中,没有采用 ModelAndView 对象,而是直接返回视图名字符串"login_success",并将数据通过设置 session 域对象属性的方式(session.setAttribute("username",username))传递给前端。

⑥前端控制器 DispatcherServlet 接收到视图名字符串,则调用视图解析器解析。

⑦解析出完整的视图文件名"webapp/login_success.jsp",返回给前端控制器 DispatcherServlet。

⑧前端控制器 DispatcherServlet 找到视图文件,并根据传递回来的数据对视图进行渲染。

⑨将视图渲染结果返回给客户端浏览器显示。

9.3 SpringMVC 详细介绍

通过上一节的介绍,我们对 SpringMVC 框架有了一个初步的认识,接下来再进行详细的介绍,学习前端控制器 DispatcherServlet、配置文件、注解的设置,以及参数的传递。

9.3.1 SpringMVC 框架的前端控制器

SpringMVC 框架的核心是前端控制器 DispatcherServlet,其实现类是 org.springframework。

web.servlet.DispatcherServlet 类,需要在项目配置文件 web.xml 中进行配置,代码如下:

```xml
<!--配置前端控制器-->
<servlet>
    <servlet-name>springmvc</servlet-name>
    <servlet-class>org.springframework.web.servlet.DispatcherServlet</servlet-class>
    <!--设置初始化时加载指定位置的 springmvc 框架的配置文件-->
    <init-param>
        <param-name>contextConfigLocation</param-name>
        <param-value>classpath:springmvc-config.xml</param-value>
    </init-param>
    <!--"1"表示容器在启动时立即加载 servlet-->
    <load-on-startup>1</load-on-startup>
</servlet>
<servlet-mapping>
    <servlet-name>springmvc</servlet-name>
    <url-pattern>/</url-pattern>
</servlet-mapping>
```

配置时通过<url-pattern>设置要拦截的地址,还要指定 SpringMVC 框架的配置文件,初始化时会加载这个配置文件。SpringMVC 框架配置文件默认文件名"springmvc-servlet.xml",默认保存位置是在"WEB-INF"目录中,其中"springmvc"是配置前端控制器时设置的 Servlet 名称。SpringMVC 框架的配置文件如果没有按默认的命名方式和位置保存,那么就必须在 web.xml 文件中使用 contextConfigLocation 进行定位,"classpath:springmvc-config.

xml"表示配置文件是位于在 src 目录中的 springmvc-config.xml 文件。

9.3.2　SpringMVC 框架配置文件

SpringMVC 框架配置文件代码如下。

```xml
1   <?xml version="1.0" encoding="UTF-8"?>
2   <beans xmlns="http://www.springframework.org/schema/beans"
3   xmlns:xsi="http://www.w3.org/2001/XMLSchema-instance"
4   xmlns:context="http://www.springframework.org/schema/context"
5   xmlns:mvc="http://www.springframework.org/schema/mvc"
6   xsi:schemaLocation="http://www.springframework.org/schema/beans
7   http://www.springframework.org/schema/beans/spring-beans-4.3.xsd
8   http://www.springframework.org/schema/context
9   http://www.springframework.org/schema/context/spring-context-4.3.xsd
10  http://www.springframework.org/schema/mvc
11  http://www.springframework.org/schema/mvc/spring-mvc-4.3.xsd">
12  <!--支持注解-->
13  <mvc:annotation-driven/>
14  <!--注解扫描-->
15  <context:component-scan base-package="controller"/>
16  <!--配置视图解析器-->
17  <bean id="viewResolver" 18 class="org.springframework.web.servlet.view.InternalResourceViewRes19olver">
20      <!--设置前缀-->
21      <property name="prefix" value="/"/>
22      <!--设置后缀-->
23      <property name="suffix" value=".jsp"/>
24  </bean>
25  </beans>
```

其中,2~11 行代码是对 beans 节点进行命名空间的声明及 mvc、context 节点的声明,13 行是声明注解支持、15 行指定注解扫描的包,17~24 行配置视图解析器,并设置了前缀和后缀。在控制器类的方法中,当返回视图名字符串时,经过视图解析器,就会在视图名前面加上前缀,后面加上后缀,构成完整的视图文件名。

当要使用其他功能的组件时,需要在配置文件中进行配置,如需要实现文件上传,那么就要配置文件上传解析器 MultipartResolver,这将在后面章节进行介绍。

9.3.3　SpringMVC 框架的注解

在 Spring 框架中,除了使用 XML 文件对各层组件进行配置外,还可以采用一种高效的方式进行配置,那就是采用注解的形式,只需要在相应的单词前加上@,就可以对各层组件进行配置,这需要 spring-aop-5.3.6.jar 包的支持,要先将这个包导入项目的 lib 目录中。下面先列出各层组件对应的注解单词。

@Component：表示是一个组件，且可以是任何层的组件。

@Repository：表示是数据访问层的组件。

@Service：表示是业务逻辑层的组件。

@Controller：表示是控制层的组件。

1. @Controlller 注解

在前面 springmvc_login 的例子中，编写的控制器类 LoginController 通过加上 @Controller 注解表明这是一个控制器类，再结合 springmvc_servlet.xml 配置文件中配置的自动扫描机制，指定要扫描 controller 包的代码：<context:component-scan base-package="controller"/>。示例代码如下：

```java
package controller;
import Javax.servlet.http.HttpSession;
import org.springframework.stereotype.Controller;
import org.springframework.web.bind.annotation.RequestMapping;
@Controller
public class LoginController{
    @RequestMapping(value="/login.do",method="post")
    public String login(String username,String password,HttpSession session){
        if(username.equals("admin")&&password.equals("123456")){
            session.setAttribute("username",username);
            return "login_success";
        }else{
            return "login_failure";
        }
    }
}
```

2. @RequestMapping

@RequestMapping 是注解请求映射名称，可以加在控制器类的上面，也可以加在控制器类方法的上面，注解这个类或方法的请求映射名称，同时还可以设置请求方式。例如在 login() 方法上面加上 @RequestMapping(value="login.do",method="GET")，表示当以 GET 方式请求 login.do 时，对应的方法是 login()。也可以设置请求映射名称为 login 或 login.action，当用户请求 login.do 时，通过处理器映射器 HandlerMapping 找到对应的 login() 方法，同时限定请求的方式是 GET。如果是直接通过浏览器地址栏访问的，请求方式是 GET；如果是表单提交的，根据提交的方式既可以是 GET，也可以是 POST。除了 GET 和 POST 这两个比较常用的方式外，还可以是 HEAD、OPTIONS、PUT、PATCH、DELETE 和 TRACE 这些方式，需要的话可以查阅有关资料。

在一个控制器类中，可以定义多个方法，对每个方法注解请求映射名称，可以直接注解在每个方法的上面，请求时直接映射到这个方法，如 http://localhost:8080/queryAllUsers.do、 http://localhost:8080/queryUserById.do。也可以对控制器类进行注解,然后再对每个方法进行注解,则请求映射就变为 http://localhost:8080/user/queryAllUsers.do、http://localhost:8080/user/queryUserById.do,控制器类注解就作为请求映射的一个路径。示例代表如下:

```
package controller;
import Javax.servlet.http.HttpSession;
import org.springframework.stereotype.Controller;
import org.springframework.web.bind.annotation.RequestMapping;
@Controller
@RequestMapping(value = "user")
public class userController{
    @RequestMapping(value = "queryAllUsers.do",method = "GET")
    public String queryAllUsers(){
    ...
    }
    @RequestMapping(value = "queryUserById.do",method = "GET")
    Public String queryUserById(int id){
    ...
    }
    @RequestMapping(value = "insert.do",method = "POST")
    Public String insert(User user){
    ...
    }
    ...
}
```

9.3.4 SpringMVC 参数传递

SpringMVC 参数传递指的是数据从前端页面传递到控制器或数据从控制器传递回前端页面。

1. 前端页面传递到控制器

从前端页面传递到控制器有两种情形:一种情形是参数直接附加在请求的 URL 后面,通过问号来携带参数,如 http://localhost:8080/bookstore/queryUserById?id=3;另一种是通过页面表单提交数据,如注册表单。在控制器类的方法中,通过定义参数来接收前端页面传递过来的数据,数据的类型可以是 Boolean、Int、String 等基本数据类型,也可以是集合类型、HttpServletRequest 类型或 HttpServletResponse 类型。

下面介绍参数附加在请求的 URL 后面的是如何接收这种参数的。

编写的控制器类 UserController.java,代码如下:

```
@Controller
public class UserController{
    @RequestMapping(value = "queryUserById1.do")
```

```java
public String queryUserById1(HttpServletRequest request){
    String id = request.getParameter("id");
    System.out.println("id: " +id);
    return "success";
}
@RequestMapping(value = "queryUserById2.do")
public String queryUserById2(int id){
    System.out.println("id: " +id);
    return "success";
}
```

当发出请求的 URL 是 http://localhost:8080/springmvc_parameter/queryUserById1.do? id = 3 时,请求映射到控制器类 UserController 中的 queryUserById1() 方法,采用定义 HttpServletRequest 类型参数的这种形式来接收 id 值,执行的结果如图 9.9 所示。当发出请求的 URL 是 http://localhost:8080/springmvc_parameter/queryUserById2.do? id = 3 时,请求映射到控制器类 UserController 中的 queryUserById2() 方法,采用普通的数据类型 Int 来接收 id 的值,id 也自动转换为 Int 类型。

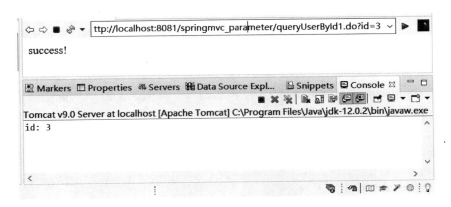

图 9.9 参数类型为 **HttpServletRequest** 的执行结果

下面再通过用户注册功能的实现介绍前端页面如何将数据传递到控制器。
前端页面 register.jsp 的代码如下:

```
<form name = "form1" method = "post" action = "reg1.do" >
    username: <input type = "text" name = "username" > <br/ >
    sex: <input type = "radio" name = "sex" value = "female" >female
        <input type = "radio" name = "sex" value = "male" >male<br/ >
    age: <input type = "text" name = "age" > <br/ >
    password: <input type = "password" name = "password" > <br/ >
    hobby: <input type = "checkbox" name = "hobbies" value = "reading" >reading
        <input type = "checkbox" name = "hobbies" value = "shopping" >shopping
```

```html
<input type = "checkbox" name = "hobbies" value = "sport">sport<br>
<input type = "submit" value = "reg">
</form>
```

控制器类 UserController.java 的代码如下:

```java
@Controller
public class UserController {
    @RequestMapping(value = "reg1.do")
    public String reg1(String username,String sex,int age,String password,String[] hobbies,Model model){
        System.out.println("username:" + username);
        System.out.println("sex:" + sex);
        System.out.println("age:" + age);
        System.out.println("password:" + password);
        System.out.println("hobbies:");
        for(String hobby:hobbies){
            System.out.println(hobby);
        }
        return "success";
    }
}
```

在控制器类的 reg1() 方法中,定义参数 username、sex、age、password、hobbies,只要名称和表单元素名相同,就可以接收前端页面的数据。当参数名称和表单元素名不一致时,可以加上@RequestParam 进行注解,指定其接收的是哪个元素名。注册功能的运行结果如图 9.10、图 9.11 和图 9.12 所示。

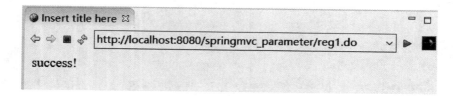

图 9.10 注册表单运行界面

图 9.11 注册成功界面

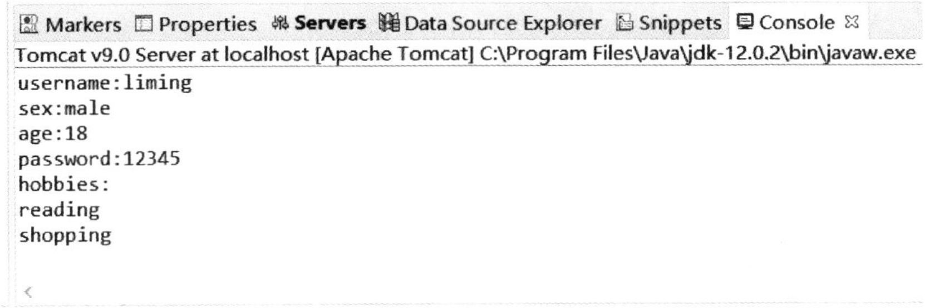

```
username:liming
sex:male
age:18
password:12345
hobbies:
reading
shopping
```

图 9.12 注册时在控制台输出的信息界面

在定义参数时,还可以根据需要指定其数据类型,这时 SpringMVC 框架会自动进行类型转换。一般参数数目不超过 5 个,可以逐个罗列出来。如果参数太多,需要将其进行封装,然后再使用对象进行传参。如下面的代码,将数据封装为 User 对象,对象的属性名与表单元素名相同。User 对象是一个 POJO 对象,即简单的 Java 对象,只包括属性和相应的 setter 和 getter 方法,没有实现其他功能的方法。

封装数据 User.java 文件:

```java
package vo;
import java.util.Arrays;
import java.util.List;
public class User {
    private String username;
    private String sex;
    private int age;
    private String password;
    private String[] hobbies;
    public String getUsername() {
        return username;
    }
    public void setUsername(String username) {
        this.username = username;
    }
    public String getSex() {
        return sex;
    }
    public void setSex(String sex) {
        this.sex = sex;
    }
    public int getAge() {
        return age;
    }
    public void setAge(int age) {
```

```java
        this.age = age;
    }
    public String getPassword() {
        return password;
    }
    public void setPassword(String password) {
        this.password = password;
    }
    public String[] getHobbies() {
        return hobbies;
    }
    public void setHobbies(String[] hobbies) {
        this.hobbies = hobbies;
    }
    @Override
    public String toString() {
        return "User [username = " + username + ", sex = " + sex + ", age = " + age + ", password = " + password + ", hobbies = " + Arrays.toString(hobbies) + "]";
    }
}
```

编写控制器类的 reg2( )方法,代码如下,使用了 User 对象作为参数,接收前端表单中跟 User 对象属性名相同的表单元素的值,通过"System. out. println(user) ;"语句将 user 对象在控制台中输出,可以看到各个属性的值。

```java
@Controller
public class userController {
    @RequestMapping(value = "reg2.do")
    public String reg2(User user,Model model){
        System.out.println(user);
        model.addAttribute("user",user);
        return "success";
    }
    @RequestMapping(value = "toReg.do")
    public String toReg(){
        return "register";
    }
}
```

封装为 User 对象进行传递的运行结果如图 9.13 所示。

2. 从控制器传递回前端页面

在控制器类的方法中,如果有数据需要传递回前端页面,那么有以下几种方式:

①如果控制器类的方法返回类型是 ModelAndView,这时既可以添加模型数据,也可以

设置视图名。示例代码如下：

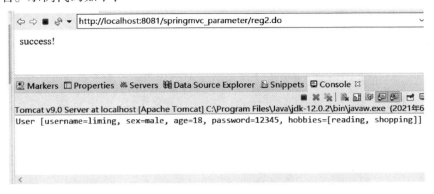

图 9.13 封装为 User 对象进行传参的运行结果

```
@RequestMapping(value = "reg3.do")
   public ModelAndView reg3(User user){
      ModelAndView mv = new ModelAndView();
      mv.addObject("RegisterUser",user);
      mv.setViewName("register_success");
      return mv;
   }
```

前端页面 Register_success.jsp，使用 EL 表达式的形式来显示传递过来的数据。示例代码如下：

```
<body>
   register info:<br>
   username:${RegisterUser.username} <br>
   sex:${RegisterUser.sex} <br>
   age:${RegisterUser.age} <br>
   password:${RegisterUser.password} <br>
   hobbies:${RegisterUser.hobbies[0]} <br>
           ${RegisterUser.hobbies[1]} <br>
</body>
```

控制器通过 ModelAndView 将数据传递回前端页面的执行结果如图 9.14 所示。

图 9.14 数据从控制器传递回前端页面的执行结果界面

②如果控制器类的方法返回类型是 String，可以通过定义 Model 参数的形式来传递数据。示例代码如下：

```
@RequestMapping(value = "reg4.do")
   public String reg4(User user,Model model){
       model.addAttribute("RegisterUser",user);
       return "register_success";
   }
```

前端页面 Register_success.jsp，也是使用 EL 表达式的形式来显示传递过来的数据。示例代码如下：

```
<body>
    register info:<br>
    username:${RegisterUser.username} <br>
    sex:${RegisterUser.sex} <br>
    age:${RegisterUser.age} <br>
    password:${RegisterUser.password} <br>
    hobbies:${RegisterUser.hobbies[0]} <br>
           ${RegisterUser.hobbies[1]} <br>
</body>
```

③借助 HttpSession 类型将数据由控制器传递回前端。实现登录功能时，登录完成后可以将登录的用户名等信息利用 session.setAttribute("loginUser",user)，将数据保存为一个 session 域变量属性，然后在前端页面显示出来。

9.4 SpringMVC 数据校验

在 Web 应用系统中，经常要对用户填写的数据进行校验，校验的操作既可以由前端执行，以便减少后台的负担，也可以在后台执行。数据校验在后台执行时有两种方式，一种是利用 Hibernate 框架的校验器实现校验，另一种是利用 Spring 框架提供的 validator 接口实现校验。在这里，介绍第一种校验方式，即利用 Hibernate 框架的校验器实现校验。

利用 Hibernate 框架的校验器进行数据校验需要以下步骤。

9.4.1 导入三个包

```
hibernate-validator-4.2.0.Final.jar
slf4j-api-1.6.1.jar
validation-api-1.0.0.GA.jar
```

slf4j-api-1.6.1.jar 和 validation-api-1.0.0.GA.jar 是 hibernate-validator-4.2.0.Final.jar 的依赖包。slf4j-api.jar 包为各种 loging APIs 提供一个简单、统一的接口，使用时需要再给它提供实现这个接口的日志包，如 log4j．common logging 等。

9.4.2　在 springmvc-servlet.xml 文件中配置 validator

```xml
<!-- 配置校验器 -->
<bean id="validator" class="org.springframework.validation.beanvalidation.LocalValidatorFactoryBean">
    <!-- hibernate 校验器 -->
    <property name="providerClass" value="org.hibernate.validator.HibernateValidator"/>
    <!-- 指定校验使用的资源文件,在文件中配置校验错误信息,如果不指定,则默认在 classpath 下的 ValidationMessages.properties -->
    <property name="validationMessageSource" ref="messageSource"/>
```

9.4.3　在数据模型中对各项数据进行校验

1. 长度检查

@Size(min=, max=):验证对象(Array, Collection, Map, String)长度是否在给定的范围之内。

@Length(min=, max=):验证字符串的长度是否在给定的范围之内,包括两端。

2. 日期检查

@Past:验证 Date 和 Calendar 对象是否在当前时间之前。

@Future:验证 Date 和 Calendar 对象是否在当前时间之后。

3. 数值检查

@Min:验证 Number 和 String 对象是否大于等于指定的值。

@Max:验证 Number 和 String 对象是否小等于指定的值。

4. 正则表达式校验规则

@Pattern:验证 String 对象是否符合正则表达式的规则。

5. 常用的校验规则

@Email:验证是否是邮件地址,如果为 null,不进行验证,直接通过验证。

@CreditCardNumber:信用卡验证。

@Range(min=, max=):检查数字是否介于 min 和 max 之间。

9.4.4　在控制器中指明哪个参数需要校验,绑定校验结果

①@Valid:校验注解,注解哪个参数需要校验。

②BindingResult:绑定校验结果,可以将结果打印出来,知道校验得怎样。

③hasErrors():判断是否有错,再决定转向哪个视图。

注册页面 register.jsp,在这个页面中校验用户填写的各项数据是否符合要求。示例代码如下:

```
<body>
    <form method="post" action="register.do">
```

```html
<table border=0>
<tr>
    <td>username:</td>
    <td><input type="text" name="username"><span><font color="red">用户名必须为字母或数字,长度在6~8个字符</font></span></td>
</tr>
<tr>
    <td>password:</td>
    <td><input type="password" name="password"><span><font color="red">密码必须是6~18位数字</font></span></td>
</tr>
<tr>
    <td>email:</td>
    <td><input type="text" name="email"><span><font color="red">要符合email的格式</font></span></td>
</tr>
<tr>
    <td>age:</td>
    <td><input type="text" name="age"><span><font color="red">年龄不能为空</font></span><td>
</tr>
<tr>
    <td colspan="2"><input type="submit" value="register"></td>
</tr>
</table>
</form>
```

对 User 对象的属性定义校验规则:

```java
public class User {
    @Length(min=6,max=20,message="用户名的长度在6~8个字符")
    @Pattern(regexp="^[a-zA-Z0-9]+$",message="用户名必须为字母或数字")
    private String username;
    @Pattern(regexp="^[0-9]{6,18}$",message="密码必须是6~18位数字")
    private String password;
    @Email(message="email地址格式不正确")
    private String email;
    @NotNull(message="年龄不能为空")
    @Range(min=1,max=99,message="年龄在1~99")
    private Integer age;
    ...
}
```

在控制器类 RegisterController.java 中，@Valid 注解 User 对象需要校验，BindingResult 绑定校验结果，如果有错，则转到显示错误信息的页面 register_fail.jsp。代码如下：

```java
@Controller
public class RegisterController {
    @RequestMapping("/register.do")
    public String register(@Valid User user, BindingResult br) {
        if(br.hasErrors()) {
            return "register_fail";
        }else{
            return "register_success";
        }
    }
}
```

显示校验错误信息的页面 register_fail.jsp，使用 Spring 框架的表单标签，form 标签的 commandName 属性设置进行校验的对象是"user"，errors 标签的 path 属性设置有校验错误的对象属性，如"username"。

```jsp
<%@ taglib prefix="s" uri="http://www.springframework.org/tags/form" %>
<body>
register failure!
    <s:form commandName="user">
    <s:errors path="username"></s:errors><br>
    <s:errors path="password"></s:errors><br>
    <s:errors path="email"></s:errors><br>
    <s:errors path="age"></s:errors><br>
    </s:form>
</body>
```

9.5 SpringMVC 文件上传

文件上传是 Web 应用系统的一个基本功能，注册时可以上传照片，添加信息时可以按要求将文件上传到服务器。

Spring 框架通过 MutipartResolver 接口的实现类来实现文件上传，从 Spring 3.1 开始，提供了两个实现类：StandardServletMultipartResolver 和 CommonsMultipartResolver。StandardServletMultipartResolver 类使用了 Servlet 3.0 所提供的对 multipart 请求的支持，不依赖第三方组件；而 CommonsMultipartResolver 使用 Apach 的 Commons FileUpload 组件来解析 multipart 请求。下面介绍如何使用 CommonsMultipartResolver 类实现文件上传。

使用 CommonsMultipartResolver 实现类实现文件上传，需要以下几个步骤。

9.5.1 在 springmvc-servlet.xml 文件中配置 MultipartResolver 组件

```xml
<!-- 支持文件上传 -->
<bean id="multipartResolver" class="org.springframework.web.multipart.commons.CommonsMultipartResolver">
    <!-- 设置默认的编码格式 -->
    <property name="defaultEncoding" value="utf-8"/>
    <!-- 上传文件大小上限,单位是字节 -->
    <property name="maxUploadSize" value="5242880"></property>
</bean>
```

9.5.2 导入两个 jar 包

```
commons-upload.jar
commons-io.jar
```

将实现文件上传的 commons-upload.jar 和 commons-io.jar 两个包复制到 lib 目录下。

9.5.3 编写上传文件的表单

代码如下:

```html
<body>
<form name="form1" action="upload.do" method="post" enctype="multipart/form-data">
请选择要上传的文件:<input type="file" name="uploadFile">
<input type="submit" value="上传">
</form>
</body>
```

\<input type="file" name="uploadFile"\>使用 file 表单元素供用户选择文件,要实现文件上传,表单的提交方式必须设置为 method="post",同时要设置 enctype 的属性值。enctype 设置编码类型,当设置为"multipart/form-data",表示以二进制形式传输数据,适合上传文件。

9.5.4 编写实现上传功能的控制器类

示例代码如下:

```java
@Controller
public class UploadController{
    @RequestMapping(value="upload.do")
    public String upload(@RequestParam("uploadFile") MultipartFile file,HttpServletRequest request,HttpSession session)throws IOException{
        System.out.println("name:"+file.getName());
        System.out.println("size:"+file.getSize());
```

```
        System.out.println("oname:"+file.getOriginalFilename());
        session.setAttribute("filename",file.getOriginalFilename());
        System.out.println("contentType:"+file.getContentType());
        String path = request.getServletContext().getRealPath("/upload");
         IOUtils.copy(file.getInputStream(), new FileOutputStream(path + "/" +
file.getOriginalFilename()));
        return "success";
    }
}
```

在编写控制器类的方法 upload 时,关键点是要将前端页面文件 upload.jsp 表单中的 uploadFile 元素设置为 MultipartFile 类型,这时可以利用 MultipartFile 提供的方法获取到上传文件的信息,如表 9.1 所示。

表 9.1　MultipartFile 的方法名及功能

方法名	功能
getName( )	得到表单中 file 元素的名称
getSize( )	得到上传文件的大小
getOriginalFilename( )	得到上传文件的文件名
getContentType( )	得到上传文件的文件类型
transferTo( )	将上传文件保存在服务器指定目录下
IOUtil.copy	将指定文件复制到服务器的指定目录下,并命名

9.5.5　运行项目

将项目部署到服务器,在浏览器中输入地址,访问上传文件的页面 upload.jsp,如图 9.15 所示。选择文件,然后单击上传按钮,由 upload.do 进行处理,若上传成功,则转到 success.jsp 页面,显示"upload success!",如图 9.16 所示。

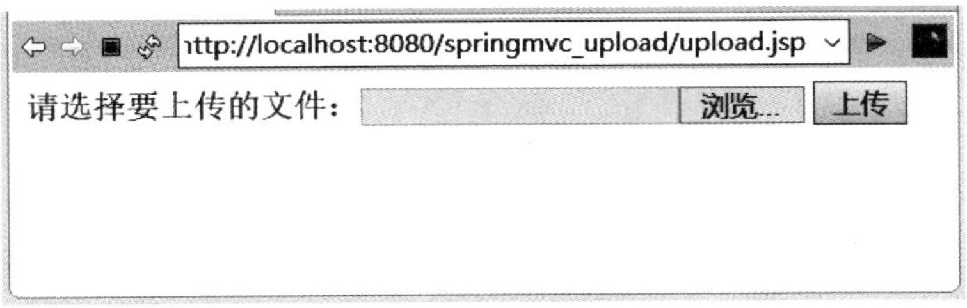

图 9.15　上传文件的界面

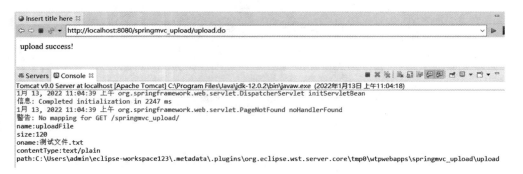

图 9.16　上传成功的界面

保存上传文件的目录如图 9.17 所示,在项目根目录下的 upload 目录中,可以查看到已上传的文件"测试文件.txt",如图 9.18 所示。

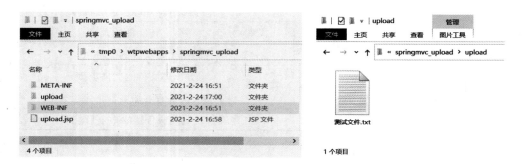

图 9.17　保存上传文件的目录　　　　图 9.18　已上传的文件

习　　题

使用 SpringMVC 实现用户注册功能,并对注册数据进行校验。注册用户包含三项信息,即用户名、密码、邮箱。注册成功后再上传照片。

第 10 章 Mybatis 框架

【本章内容】

- 10.1 Mybatis 框架概述
- 10.2 Mybatis 框架详细介绍
- 10.3 Mybatis 框架 SQL 映射文件

10.1 Mybatis 框架概述

10.1.1 什么是 Mybatis 框架

Mybatis 框架(前身是 Ibatis)是一个支持普通 SQL 查询、存储过程及高级映射的持久化层框架,是一个 ORM 框架。ORM(Object Relation Mapping)即对象关系映射。在 Java EE 技术体系中,因为 Java 语言是纯面向对象编程语言,而数据库一般是关系型数据库,将对关系型数据库表的增删改查的操作映射为 Java 对象的增删改查的方法称为对象关系映射。

相比较另一个 ORM 框架 Hibernate,Mybatis 框架是一个半自动的映射框架,在映射时,还需要手动编写 SQL 语句,因此,可以根据业务需求灵活地编写 SQL 语句,从而提升 Mybatis 框架的性能,对于一些复杂的和需要优化性能的项目来说,Mybatis 框架比 Hibernate 框架更适合。

Mybatis 框架可以在 https://github.com/mybatis/mybatis-3/releases 下载,其中包含一个核心包 mybatis-3.5.5.jar 和使用手册。

10.1.2 Mybatis 的工作流程

Mybatis 框架底层封装了 JDBC(Java DataBase Connectivity),下面通过与使用 JDBC 操作数据库的流程做一下对比,介绍 Mybatis 的工作流程,如图 10.1 所示。

Mybatis 框架操作数据库的流程如下:

①加载 Mybatis 框架配置文件 mybatis-config.xml,并读取映射文件 mapper.xml。

②根据配置文件创建会话工厂 SqlSessionFactory。

③由会话工厂 SqlSessionFactory 创建会话对象 SqlSession,一个会话对象相当于一次数据库连接。

④一个会话对象 SqlSession 包含有一个 Executor 对象,用来执行 Query 或 Update 操作,找到映射文件 mapper.xml 里面定义的 Statement ID,执行对应的 SQL 语句。

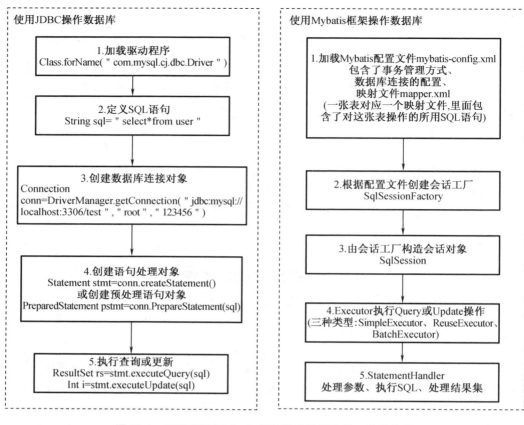

图 10.1　JDBC 和 Mybatis 框架操作数据库的工作流程图

⑤执行时会创建 StatementHandler，根据 StatementType 决定 StatementHandler 的类型是 STATEMENT、PREPARED 还是 CALLABLE，默认是 PREPARED。StatementHandler 里面包含了处理参数的 ParmeterHandler 和处理结果集的 ResultSetHandler。用 PrepareStatement( ) 方法对语句进行预编译，处理参数，执行 PreparedStatement 的 execute( ) 方法，最后通过 ResultSetHandler 处理结果集，将输出结果映射为 Java 对象。

10.1.3　使用 Mybatis 框架操作数据库的基本步骤

①创建项目，将 Mybatis 框架 jar 包和 MySQL 数据库驱动包复制到 lib 目录中。
②在 src 目录中创建 Mybatis 框架的配置文件 mybatis-config.xml，该文件如下。

```
<?xml version = "1.0" encoding = "UTF-8"?>
<!DOCTYPE configuration
    PUBLIC "-//mybatis.org//DTD Config 3.0//EN"
    "http://mybatis.org/dtd/mybatis-3-config.dtd">
<configuration>
    <environments default = "development">
        <environment id = "development">
            <transactionManager type = "JDBC"/>
```

```xml
        <dataSource type="POOLED">
            <property name="driver" value="com.mysql.cj.jdbc.Driver"/>
            <property name="url" value="jdbc:mysql://localhost:3306/jdbctest?useUnicode=true&characterEncoding=UTF-8&serverTimezone=Asia/Shanghai"/>
            <property name="username" value="root"/>
            <property name="password" value="123456"/>
        </dataSource>
    </environment>
  </environments>
  <mappers>
    <mapper resource="dao/UserDAO.xml"/>
  </mappers>
</configuration>
```

在配置文件中,指定了事务管理方式,设置了数据库驱动程序名、连接地址、连接用户名和密码,同时指明SQL映射文件。

③创建pojo包,在其中创建持久化类User.java,代码如下:

```java
package pojo;
public class User {
    private int id;
    private String username;
    private String password;
    public int getId() {
        return id;
    }
    public void setId(int id) {
        this.id = id;
    }
    public String getUsername() {
        return username;
    }
    public void setUsername(String username) {
        this.username = username;
    }
    public String getPassword() {
        return password;
    }
    public void setPassword(String password) {
        this.password = password;
    }
    @Override
```

```
    public String toString()
        return "User [id = " + id + ", username = " + username + ", password = " + password + "]";
    }
}
```

④创建 dao 包、UserDAO.java 接口文件和 UserDAO.xml 映射文件。

UserDAO.java 接口文件代码如下。

```
package dao;
import java.util.List;
import org.apache.ibatis.annotations.Param;
public interface UserDAO {
    public List<User> queryAll();//查询所有用户
    public User queryUserById(int id); //查询指定 ID 的用户
}
```

UserDAO.xml 映射文件代码如下。

```
<?xml version = "1.0" encoding = "UTF-8"?>
<!DOCTYPE mapper
    PUBLIC "-//mybatis.org//DTD Mapper 3.0//EN"
    "http://mybatis.org/dtd/mybatis-3-mapper.dtd">
<mapper namespace = "dao.UserDAO">
    <select id = "queryAll" resultType = "pojo.User">
        select * from user;
    </select>
    <select id = "queryUserById" resultType = "pojo.User">
        select * from user where id = #{id};
    </select>
</mapper>
```

⑤在 src 目录中,创建测试类 test.java,完成查询所有用户和根据指定 ID 查询用户的功能代码如下。

```
public class test {
    public static void main(String[] args) {
        queryAll();
        queryUserById();
    }
    public static void queryAll() {
        //读取配置文件
        InputStream is = Resources.class.getResourceAsStream("/mybatis-config.xml");
        //根据配置文件创建 SqlSessionFactory 对象
        SqlSessionFactory sf = new SqlSessionFactoryBuilder().build(is);
        //由 SqlSessionFactory 对象创建 SqlSession 对象
```

```java
        SqlSession session = sf.openSession();
        //由接口得到实现接口的映射文件
        UserDAO userDAO = session.getMapper(UserDAO.class);
        //得到查询结果
        List<User> list = userDAO.queryAll();
        //通过for循环将结果输出
        for(User u:list){
            System.out.println(u);
        }
        session.close();
    }
    public static void queryUserById(){
        InputStream is = Resources.class.getResourceAsStream("/mybatis-config.xml");
        SqlSessionFactory sf = new SqlSessionFactoryBuilder().build(is);
        SqlSession session = sf.openSession();
        UserDAO userDAO = session.getMapper(UserDAO.class);
        User user = userDAO.queryUserById(4);
        session.close();
        System.out.println(user);
    }
}
```

项目的目录结构如图 10.2 所示。

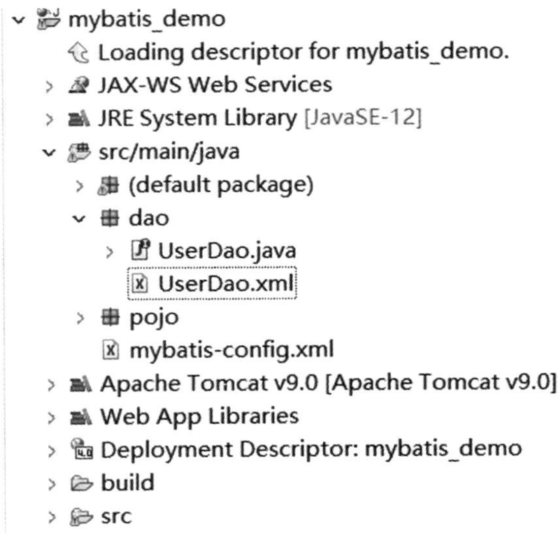

```
v 🗁 mybatis_demo
    🔍 Loading descriptor for mybatis_demo.
  > 🗁 JAX-WS Web Services
  > 🗁 JRE System Library [JavaSE-12]
  v 🗁 src/main/java
    > 🗁 (default package)
    v 🗁 dao
      > 🗎 UserDao.java
        🗎 UserDao.xml
    > 🗁 pojo
        🗎 mybatis-config.xml
  > 🗁 Apache Tomcat v9.0 [Apache Tomcat v9.0]
  > 🗁 Web App Libraries
  > 🗁 Deployment Descriptor: mybatis_demo
  > 🗁 build
  > 🗁 src
```

图 10.2 mybatis_demo 项目的目录结构图

10.2 Mybatis 框架详细介绍

10.2.1 Mybatis 框架配置文件

Mybatis 框架的配置文件 mybatis-config.xml 是一个全局配置文件,通常命名为 mybatis-config.xml,并保存在 src 目录中。从文件名能非常清楚地知道这是 Mybatis 框架的配置文件,在配置文件中设置数据库连接、事务管理方式和指明 SQL 映射文件。配置文件的根节点是 configuration,下面代码是前面演示案例的配置文件。

```
1  <?xml version = "1.0" encoding = "UTF-8"?>
2  <!DOCTYPE configuration
3   PUBLIC "-//mybatis.org//DTD Config 3.0//EN"
4   "http://mybatis.org/dtd/mybatis-3-config.dtd">
5  <configuration>
6   <environments default = "development">
7    <environment id = "development">
8     <transactionManager type = "JDBC"/>
9     <dataSource type = "POOLED">
10     <property name = "driver" value = "com.mysql.cj.jdbc.Driver"/>
11     <property name = "url"
12       value = "jdbc:mysql://localhost:3306/jdbctest?useUnicode = true&characterEncoding = UTF-8&serverTimezone = Asia/Shanghai"/>
13     <property name = "username" value = "root"/>
14     <property name = "password" value = "123456"/>
15    </dataSource>
16   </environment>
17  </environments>
18  <mappers>
19   <mapper resource = "dao/UserDAO.xml"/>
20  </mappers>
21 </configuration>
```

第 1 行代码指明 XML 的版本和支持的编码格式,第 2~4 行指明了根节点 configuration 所使用的文档类型并定义文档 mybatis-3-config.dtd,这个文档规范了根节点 configuration 及子节点的名称及约束,设置有 < properties > < settings > < typeAliases > < typeHandler > < objectFactory > < plugins > < environments > < databaseIdProvider > 和 < mappers > 9 个子节点。在这个演示案例的配置文档中,只对 < environments > 和 < mappers > 两个子节点进行了设置,这两个节点是必须设置的,关于其他子节点的设置可以参考相关的资料。

< environments > 节点用于配置数据库环境,下面可以有多个 < environment > 子节点,可以设置多个数据库环境, < environment id = "development" > 设置了这个数据库环境的

id,并作为缺省的环境,`<environments default="development">`。配置数据库环境需要配置事务管理方式和数据源,`<transactionManager type="JDBC">`设置了事务管理方式是"JDBC",表示采用 JDBC 的事务管理方式,即需要手动提交和回滚设置。另一种事务管理方式是"MANAGED",则将事务交给容器来管理。

`<dataSource>`子节点设置数据源,属性`type="POOLED"`设置了数据源采用连接池,存储一定量的连接对象,减少了创建连接和关闭连接的开销,提高了数据库访问效率,是当前流行的数据源类型。`type="UNPOOLED"`设置不采用连接池,对于那些对访问效率要求不高的小型应用项目可以采用这种方式。`type="JNDI"`设置采用外部数据源,数据源位于其他服务器,如 EJB 或应用服务器如 JBoss,采用 JNDI 方式来访问。`<dataSource>`通过`<property>`子节点设置数据库连接的四个属性,即 driver、url、username 和 password。本章节是直接设置四个属性值,在后面 SSM 框架整合时,再把这四个属性的值放在 db.properties 文件中进行设置,这样便于属性值的修改。

`<configuration>`根节点下面的另一个子节点`<mappers>`也是必须要设置的,可以有多个`<mapper>`子节点,用于指定 SQL 映射文件,有四种路径表示方式,具体如下。

①通过 resource 属性引入类路径的相对资源,通常项目采用这种方式来引入,代码如下:

```
<mappers>
    <mapper resource="com/books/mapper/UserMapper.xml"/>
</mappers>
```

②通过 url 引入本地绝对路径的资源,代码如下:

```
<mappers>
    <mapper url="file:///d:/booksmanager/mapper/UserMapper.xml"/>
</mappers>
```

③使有 class 属性引入 mapper 接口文件,接口文件必须和 mapper 映射文件在同一目录中,并且名称相同,代码如下:

```
<mappers>
    <mapper class="com.books.mapper.UserMapper"/>
</mappers>
```

④使用包名批量引入,通过 name 属性指明映射文件所在的包名,代码如下:

```
<mappers>
    <!--将指定包中的所有映射文件全部引入-->
    <package name="com.books.mapper"/>
</mappers>
```

10.2.2 SqlSessionFactory 对象

SqlSessionFactory 是 Mybatis 框架的一个核心对象,是由 SqlSessionFactoryBuilder 构造器根据框架的配置文件构造得到的,由它再创建 SqlSession 对象。构造一个 SqlSessionFactory 对象会消耗很多资源,所以通常采用单例模式,一个应用只创建一个 SqlSessionFactory 对象,这个对象一旦创建,在整个应用程序运行期间一直存在。SqlSessionFactory 对象可以被多个线程调用以创建 SqlSession 对象。

10.2.3 SqlSession 对象

SqlSession 是 Mybatis 框架的另一个核心对象,由 SqlSessionFactory 对象创建,是应用程序与持久化类之间执行交互操作的单线程对象。Mybatis 框架对数据库执行增删改查操作是由 SqlSession 对象调用底层的 Executor 对象执行的,SQL 映射文件经过解析后再将传入的参数进行映射,由 Executor 对象再调用 StatementHandler 等对象对数据库进行增删改查操作。SqlSession 对象只能在一次请求或一个方法体中创建,使用完 SqlSession 对象后要及时关闭。SqlSession 对象提供多个方法用于执行增删改查操作,如图 10.3 所示。

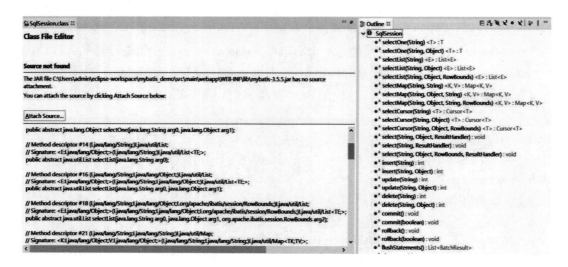

图 10.3 SqlSession 对象的增删改查方法

10.2.4 SqlSession 执行数据库操作的两种方式

1. 直接调用 SqlSession 对象的增删改查方法

以前面演示案例执行的查询所有用户和查询指定 ID 用户的操作为例,代码如下。

```
public static void queryAll(){    //查询所有用户
    InputStream is = Resources.class.getResourceAsStream("/mybatis-config.xml");
    SqlSessionFactory sf = new SqlSessionFactoryBuilder().build(is);
    SqlSession session = sf.openSession();
    List<User> list = session.selectList("dao.UserDAO.queryAll");
    for(User u:list){
        System.out.println(u);
    }
    session.close();
}
```

查询所有用户使用的是 SqlSession 对象的 selectList(String statement)方法,参数 statement 的值由 UserDAO.xml 映射文件中 <mapper> 节点的 namespace 属性值"dao.UserDAO" + 查询所有用户的 <select> 节点的 id 值"queryAll"两部分构成,其 <select> 节点的代码如下。

```
<select id = "queryAll" resultType = "pojo.User" >
    select * from user;
</select>
public static void queryUserById(){    //查询指定 ID 的用户
    InputStream is = Resources.class.getResourceAsStream("/mybatis-config.xml");
    SqlSessionFactory sf = new SqlSessionFactoryBuilder().build(is);
    SqlSession session = sf.openSession();
    User user = session.selectOne("dao.UserDAO.queryUserById",2)
    session.close();
    System.out.println(user);
}
```

查询指定 ID 用户使用的是 SqlSession 对象的 selectOne(String statement,Object parameter)方法,参数 statement 的值由 UserDAO.xml 映射文件中 <mapper> 节点的 namespace 属性值"dao.UserDAO" + 查询指定 ID 用户的 <select> 节点的 id 属性值"queryUserById"两部分构成;参数 parameter 设置要查询的用户的 ID,其 <select> 节点的代码如下,这时就会将 session.selectOne()方法中的 parameter 参数值"2"传给 SQL 映射文件中的 id。

```
<select id = "queryUserById" resultType = "pojo.User" >
    select * from user where id = #{id};
</select>
```

2. 演示案例采用的方式

Mybatis 后期的版本提供了第二种方式,也就是演示案例所采用的方式,即在同一个包中,定义一个接口文件 UserDAO.java 和一个映射文件 UserDAO.xml,两个文件的文件名相同,接口文件中定义实现增删改查的方法,在映射文件中编写这些方法的 SQL 语句,其 SQL 语句的节点的 id 值跟接口文件的方法名对应,所以根据方法名就能找到对应的要执行的 SQL 语句,现在主要是采用这种方式。

10.3 Mybatis 框架 SQL 映射文件

10.3.1 基本增删改查功能的 SQL 映射文件

使用 Mybatis 框架实现持久化层的操作,需要开发人员编写 SQL 映射文件,即编写对数据库表的增删改查操作的 SQL 语句,Mybatis 框架提供 <insert>、<delete>、<update> 和 <select> 节点用于编写增删改查的 SQL 语句,这些节点也提供一些属性供开发人员进行设置。下面就前面演示案例,把实现基本的增删改查功能的映射文件展示出来,并就各节点及其属性做详细介绍。

1. 接口文件及 SQL 映射文件

在编写 SQL 映射文件 UserDAO.xml 之前,要先编写接口文件 UserDAO.java。

UserDAO.java 接口文件:

```java
package dao;
import java.util.List;
import org.apache.ibatis.annotations.Param;
import pojo.User;
public interface UserDAO {
    public List<User> queryAll();
    public int insert(User user);
    public int updateUser(User user);
    public int delete(int id);
    public User queryUserById(int id);
}
```

UserDAO.xml 映射文件:

```xml
<?xml version="1.0" encoding="UTF-8"?>
<!DOCTYPE mapper
    PUBLIC "-//mybatis.org//DTD Mapper 3.0//EN"
    "http://mybatis.org/dtd/mybatis-3-mapper.dtd">
<mapper namespace="dao.UserDAO">
    <!--查询所有用户-->
    <select id="queryAll" resultType="pojo.User">
        select * from user;
    </select>
    <!--加用户-->
    <insert id="insert">
        insert into user(username,password) values(#{username},#{password});
    </insert>
    <!--新用户-->
    <update id="updateUser">
        update user set username=#{username},password=#{password} where id=#{id};
    </update>
    <!--除指定ID的用户-->
    <delete id="delete">
        delete from user where id=#{id};
    </delete>
    <!--查询指定ID的用户-->
    <select id="queryUserById" resultType="pojo.User">
        select * from user where id=#{id};
    </select>
</mapper>
```

第10章 Mybatis框架

映射文件的根节点是 < mapper namespace = "dao.UserDAO" >，其 namespace 属性用于设置名称空间，名称空间设置值为接口文件所在的包 + 文件名，不包括后缀，这个例子中设置为"dao.UserDAO"，通常我们也是将映射文件放在这个包中，名称跟接口文件名相同，如果不同，要注意名称空间设置的是接口文件，而不是映射文件。

2. 接口文件的方法和映射文件的 SQL 语句之间的调用关系及参数传递

< mapper > 子节点及其功能说明如表 10.1 所示。

表 10.1 < mapper > 子节点及其功能说明

子节点名称	功能说明
< select >	映射 select 语句，实现查询功能
< insert >	映射 insert 语句，实现添加功能
< update >	映射 update 语句，实现更新功能
< delete >	映射 delete 语句，实现删除功能

在实现持久化层时，由上一层调用接口文件中的方法，然后在映射文件中查找跟方法名称相同的节点，执行相应的 SQL 语句，整个过程通过图 10.4 来进行展示。

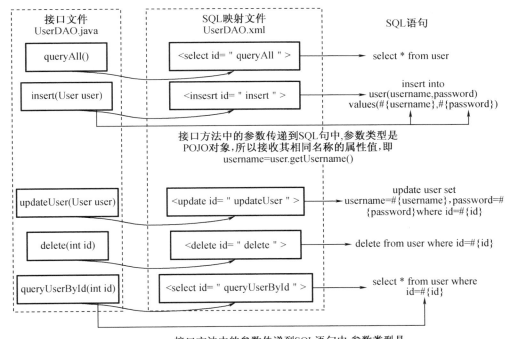

图 10.4 接口文件中的方法和映射文件中的 SQL 语句之间的调用关系及参数传递

在参数传递过程中，如果传入参数是 POJO 类型的，则接收 POJO 对象的同名属性值；如果传入参数是普通数据类型的，如 int id、String orderId，则接收其同名的参数；如果传入两个

以上普通类型参数,则在接口文件的方法中,参数必须使用@param进行注解,例如:
```
public interface UserDAO {
    //public int insert(User user);  //使用POJO类的参数
    public int insert(@Param("username") String username,@Param("password") String password);   //使用普通数据类型作为参数
}
```

3. <select>等节点常用属性的使用说明

下面接着介绍<select>、<insert>、<delete>和<update>这些节点常用的属性,它们的使用说明如表10.2所示。

表10.2 节点常用属性的使用说明

常用属性	节点名	使用说明
id	<select> <insert> <delete> <update>	SQL语句唯一的标识符,其值为接口文件UserDAO.java中对应的方法名,如<select id="queryAll">queryAll就是接口文件中的queryAll()方法
parameterType	<select> <insert> <delete> <update>	用于设置传入SQL语句的参数类型,可以是基本的数据类型或POJO类型,如果传入的参数类型是POJO类型,需要设置parameterType属性
resultType	<select>	设置select语句返回结果的类型,返回结果通常是一个POJO类型或集合类型,resultType要设置POJO类型或集合元素中的POJO类型
resultMap	<select>	指定select语句返回结果的映射,这个映射需要另外通过<resultMap>节点进行构造。resultMap用于两种情形:一种是当返回结果封装为POJO对象,但POJO对象的属性名与数据库表的字段名不一致时,可以借助resultMap来指定两者的映射关系;另一种是用于表间的关联映射,有一对一、一对多、多对一和多对多,具体在11.3.2节讲解。

4. #{}和${}的使用说明

在编写SQL映射文件时,在<select>等节点中定义的SQL语句,要传入的参数通常使用#{}来表示,如:
```
<select id="selectUserById">
    Select * from user where id=#{id};
</select>
```

#{}在底层被JDBC解析为PreparedStatement预编译语句,被当作一个整体变量,比如字符串、整型等。下面是使用预编译语句处理的示例:
```
int id=2;
String sql = "select * from user where id=?";
```

```
PreparedStatement pstm = connection.prepareStatement(sql);
pstm.setInt (1,id);
```

这里,SQL 语句只有一个参数,1 表示将整型 id 的值传入 SQL 语句中的第 1 个问号。

```
String sql = "update user set username = ?,password = ? Where id = ?"
PreparedStatement pstm = connection.prepareStatement(sql);
pstm.setString(1,user.getUsername());
pstm.setString(2,user.getPassword());
pstm.setInt (3,user.getId());
```

这里,SQL 语句有 3 个参数,1 表示 SQL 语句中的第 1 个问号,2 表示第 2 个问号,3 表示第 3 个问号,分别传入 user 对象的 3 个属性值。

在 SQL 映射文件中,<update>节点中定义的 SQL 语句,如下:

```
<update id = "updateUser" parameterType = "user" >
    update user set username = #{username},password = #{password},where id = #{id}
</update>
```

上面程序段会将传入的 user 对象的三个同名属性值传入 SQL 语句中对应的三个参数。

有些情况下,会使用到 ${},要注意两者用法的不同。

#{}:被预编译,会自动加上单引号,可以传入对象,也可以传入变量。

${}:直接被替换,用于拼接 SQL 语句,只能接收对象参数,不能接收普通参数。

如模糊查询,在 JDBC 中我们定义的模糊查询语句是:

```
select * from user where username like '%s%'
```

其中"%"是通配符,代表任意符号,"%s%"表示查询 username 字段值中包含有字符 s 的记录。

在 SQL 映射文件中,我们定义的 SQL 语句是:

```
select * from user where username like '% #{username}%'
```

因为#{}会自动加上引号,所以,如果 username 的值为 liming,这时就变成 like '% li-ming'%,这样就不符合语法了,所以需要使用 ${}来表示,即

```
select * from user where username like '% ${username}%'
```

10.3.2 实现表间关联查询的 SQL 映射文件

数据库表与表之间的关联关系有四种类型:一对一、多对一、一对多和多对多。如一个人对应一个身份证号,一个身份证号对应一个人,人与身份证号之间是一对一的关系;一个人有多个地址,一个地址对应一个人,地址和人之间是多对一的关系;反之,人和地址之间是一对多的关系;一个学生有多个老师,一个老师有多个学生,学生和老师之间就是多对多的关系,多对多实际就是双向的一对多。

在执行表间关联查询时,Mybatis 提供了关联映射,即在 SQL 映射文件中借助前面讲过的<select>节点的 resultMap 属性构造表间的关联关系。

在 Mybatis 框架中实现表间关联查询的步骤如下:

①在数据库中对相关联的表进行处理,使两个表建立关联;

②在项目中创建两个表对应的 POJO 类,在类中增加属性,使两个对象建立关联;

③在编写 SQL 映射语句时,通过<resultMap>节点构造表间的关联,节点的 association

属性和 collection 属性定义对象之间的映射关系，association 属性用于一对一的映射，collection 属性用于一对多的映射。

1. 一对一的关联

一本图书对应一个 ISBN 号，一个 ISBN 号对应一本图书，它们之间是一对一的关联关系。实现表间关联查询的具体步骤如下：

①通过在 book 表中增加一个字段 isbn_id 作为外键，关联 isbn 表中的 id 字段，如图 10.5 和图 10.6 所示。

图 10.5 book 表　　　　　　　　　图 10.6 isbn 表

②创建对应的持久化对象 Book 和 Isbn，在 Book 对象中增加一个属性关联 isbn 对象，代码如下。

Book.java：

```java
package pojo;
public class Book {
    private int id;
    private String name;
    private Isbn isbn;    //通过设置属性的方式与另一个对象建立关联
    public int getId() {
        return id;
    }
    public void setId(int id) {
        this.id = id;
    }
    public String getName() {
      return bookname;
    }
    public void seName(String name) {
        this.name = name;
    }
    public Isbn getIsbn() {
        return isbn;
    }
    public void setIsbn(Isbn isbn) {
        this.isbn = isbn;
```

```
    }
    @Override
    public String toString() {
        return "Book [id = " + id + ", name = " + name + ", isbn = " + isbn + "]";
    }
}
```

Isbn.java：

```
package pojo;
public class Isbn {
    private int id;
    private String isbn;
    public int getId() {
        return id;
    }
    public void setId(int id) {
        this.id = id;
    }
    public String getIsbn() {
        return isbn;
    }
    public void setIsbn(String isbn) {
        this.isbn = isbn;
    }
    @Override
    public String toString() {
        return "Isbn [id = " + id + ", isbn = " + isbn + "]";
    }
}
```

③创建对 Book 对象操作的接口文件 BookDAO.java 和相应的 SQL 映射文件 BookDAO.xml。在 SQL 映射文件中添加 <select> 节点，编写相应的 SQL 语句，并使用 resultMap 构造映射结果。

接口文件 BookDAO.java：

```
package dao;
import java.util.List;
import pojo.Book;
public interface BookDAO {
    List<Book> queryAllBooks();
}
```

SQL 映射文件 BookDAO.xml：

```xml
<?xml version="1.0" encoding="UTF-8"?>
<!DOCTYPE mapper PUBLIC "-//mybatis.org//DTD Mapper 3.0//EN"
"http://mybatis.org/dtd/mybatis-3-mapper.dtd">
<mapper namespace="dao.BookDAO">
    <select id="queryAllBooks" resultMap="BookMap">
        select * from book as b,isbn as i where b.isbn_id=i.id;
    </select>
    <resultMap type="pojo.Book" id="BookMap">
        <id property="id" column="id"/>
        <result property="name" column="name"/>
        <association property="isbn" javaType="pojo.Isbn">
            <id property="id" column="id"/>
            <result property="isbn" column="isbn"/>
        </association>
    </resultMap>
</mapper>
```

在 <select> 节点中，编写 select 查询语句，通过 where b.isbn_id=i.id 将表 book 和表 isbn 进行关联，Mybatis 框架将查询结果自动封装成对象，因为是表间关联查询，查询结果不再是原来的 Book 对象，所以要构造查询结果。<resultMap> 节点的 type 属性指明要对哪个对象进行构造，这里是对 Book 对象进行构造，将关联的 Isbn 对象装填进来，两个对象是一对一的关系，所以使用 <association> 节点将关联的对象包含进来。<id> 节点对应数据库表中作为主键的字段，<result> 节点对应数据库表的普通字段。<id> 节点和 <result> 节点的 property 属性指的是在 Book 对象或 Isbn 对象中的属性名，column 属性指的是在 select 语句中设置的查询结果显示的字段名。

④编写测试程序。

```java
public class test {
    public static void main(String[] args) {
        //TODO Auto-generated method stub
        InputStream is = Resources.class.getResourceAsStream("/mybatis-config.xml");
        SqlSessionFactory sf = new SqlSessionFactoryBuilder().build(is);
        SqlSession sqlSession = sf.openSession();
        BookDAO bookDao = sqlSession.getMapper(BookDAO.class);
        List<Book> list = bookDao.queryAllBooks();
        for(Book b:list){
            System.out.println(b);
        }
    }
}
```

⑤运行结果如图 10.7 所示。

```
<terminated> test (5) [Java Application] C:\Program Files\Java\jdk-12.0.2\bin\javaw.exe (2021年6
Book [id=1, name=面向对象程序设计JAVA版, isbn=Isbn [id=1, isbn=9787111187646]]
Book [id=2, name=Web开发前端技术, isbn=Isbn [id=2, isbn=9787113271503]]
Book [id=3, name=网页设计教程, isbn=Isbn [id=3, isbn=9787121106057]]
```

图 10.7 一对一关联查询运行结果

对 SQL 映射文件 BookDAO. xml 进行改写,代码及运行结果如图 10.8 所示。

```
BookDAO.xml
<?xml version="1.0" encoding="UTF-8"?>
<!DOCTYPE mapper PUBLIC "-//mybatis.org//DTD Mapper 3.0//EN"
"http://mybatis.org/dtd/mybatis-3-mapper.dtd">
<mapper namespace="dao.BookDAO">
    <select id="queryAllBooks" resultMap="BookMap">
        select b.id bid,b.name bname,i.id id,i.isbn isbn from book as b,isbn as i where b.isbn_id=i.id;
    </select>
    <resultMap type="pojo.Book" id="BookMap">
        <id property="id" column="bid"/>
        <result property="name" column="bname"/>
        <association property="isbn" javaType="pojo.Isbn">
            <id property="id" column="id"/>
            <result property="isbn" column="isbn"/>
        </association>
    </resultMap>
</mapper>
```

```
Design  Source
Markers  Properties  Servers  Data Source Explorer  Snippets  Console
<terminated> test (5) [Java Application] C:\Program Files\Java\jdk-12.0.2\bin\javaw.exe (2021年6月11日 上午10:19:52 – 上午10:19:54)
Book [id=1, name=面向对象程序设计JAVA版, isbn=Isbn [id=2, isbn=9787111187646]]
Book [id=2, name=Web开发前端技术, isbn=Isbn [id=3, isbn=9787113271503]]
Book [id=3, name=网页设计教程, isbn=Isbn [id=1, isbn=9787121106057]]
```

图 10.8 改写的 SQL 映射文件及运行结果

再对 SQL 映射文件 BookDAO. xml 进行改写,代码及运行结果如图 10.9 所示。

从图 10.8 和图 10.9 我们可以清楚地看到,构造 BookMap 对象时, < id > 的 property 属性指的是对象的属性,column 属性对应的是 select 语句中设置的字段名,当 select 语句中 b. id 别名设置为 bid,这时 < id > 的 column 属性值就要跟这个一致,设置为 bid,如果不一致,那就无法接收 bid 的值。如在 select 语句中 name 别名设置为 bname,在 < result > 节点中的 column 属性值设置为 name,则运行结果显示 Book 对象中的 name = null。

对于一对一的关联关系,如图 10.5 和图 10.6 所示,实际上通常的处理是直接在 book 表中增加 isbn 字段,如图 10.10 所示,不需要建立关联。

```xml
<?xml version="1.0" encoding="UTF-8"?>
<!DOCTYPE mapper PUBLIC "-//mybatis.org//DTD Mapper 3.0//EN"
 "http://mybatis.org/dtd/mybatis-3-mapper.dtd">
<mapper namespace="dao.BookDAO">
    <select id="queryAllBooks" resultMap="BookMap">
        select b.id bid,b.name bname,i.id,i.isbn isbn from book as b,isbn as i where b.isbn_id=i.id;
    </select>
    <resultMap type="pojo.Book" id="BookMap">
        <id property="id" column="bid"/>
        <result property="name" column="name"/>
        <association property="isbn" javaType="pojo.Isbn">
            <id property="id" column="id"/>
            <result property="isbn" column="isbn"/>
        </association>
    </resultMap>
</mapper>
```

```
Book [id=1, name=null, isbn=Isbn [id=2, isbn=9787111187646]]
Book [id=2, name=null, isbn=Isbn [id=3, isbn=9787113271503]]
Book [id=3, name=null, isbn=Isbn [id=1, isbn=9787121106057]]
```

图 10.9　再次改写的 SQL 映射文件及运行结果

图 10.10　book 表和 isbn 表合并为一个表

2. 多对一的关联

一本图书对应一个图书类别，一个图书类别对应多本图书，所以图书与图书类别是多对一的关系。多对一的关联关系处理方式和一对一是一样的。

实现表间关联查询的具体步骤如下：

①通过在 book 表中增加一个字段 type_id 作为外键，关联 type 表中的 id 字段，如图 10.11 和图 10.12 所示。

图 10.11　book 表　　　　　　　　图 10.12　type 表

②创建对应的持久化对象 Book 和 Type，在 Book 对象中增加一个属性，关联 Type 对象，代码如下。

Book.java 文件:
```java
package pojo;
    public class Book {
    private int id;
    private String name;
    private Type type; //通过设置属性的方式与另一个对象建立关联
    public int getId() {
        return id;
    }
    public void setId(int id) {
        this.id = id;
    }
    public String getName() {
        return name;
    }
    public void setName(String name) {
        this.name = name;
    }
    public Type getType() {
        return type;
    }
    public void setType(Type type) {
        this.type = type;
    }
    @Override
    public String toString() {
        return "Book [id = " + id + ", name = " + name + ", type = " + type + "]";
    }
}
```

Type.java 文件:
```java
package pojo;
public class Type {
    private int id;
    private String bookType;
    public int getId() {
        return id;
    }
    public void setId(int id) {
        this.id = id;
    }
    public String getBookType() {
```

```
        return bookType;
    }
    public void setBookType(String bookType) {
        this.bookType = bookType;
    }
    @Override
    public String toString() {
        return "Type [id = " + id + ", bookType = " + bookType + "]";
    }
}
```

③创建对 Book 对象操作的接口文件 BookDAO.java,创建相应的 SQL 映射文件 BookDAO.xml。在 SQL 映射文件中添加 <select> 节点,编写相应的 SQL 语句,并构造关联映射结果。

接口文件 BookDAO.java 文件:

```
package dao;
import java.util.List;
import pojo.Book;
public interface BookDAO {
    List<Book> queryAllBooks();
}
```

SQL 映射文件 BookDAO.xml 文件:

```
<?xml version = "1.0" encoding = "UTF-8"?>
<!DOCTYPE mapper PUBLIC "-//mybatis.org//DTD Mapper 3.0//EN"
"http://mybatis.org/dtd/mybatis-3-mapper.dtd">
<mapper namespace = "dao.BookDAO">
    <select id = "queryAllBooks" resultMap = "BookMap">
        select b.id bid,b.name bookname,t.id tid,t.bookType type from book as b,type as t where b.type_id = t.id;
    </select>
    <resultMap type = "pojo.Book" id = "BookMap">
        <id property = "id" column = "bid"/>
        <result property = "name" column = "bookname"/>
        <association property = "type" JavaType = "pojo.Type">
            <id property = "id" column = "tid"/>
            <result property = "bookType" column = "type"/>
        </association>
    </resultMap>
</mapper>
```

④运行结果如图 10.13 所示。

```
Markers   Properties   Servers   Data Source Explorer   Snippets   Console
<terminated> test (2) [Java Application] C:\Program Files\Java\jdk-12.0.2\bin\javaw.exe (2021年6月11日 上午
Book [id=1, name=java面试一战到底(基础卷), type=Type [id=1, bookType=计算机与电子信息]]
Book [id=2, name=Java程序员面试笔试通关宝典, type=Type [id=1, bookType=计算机与电子信息]]
Book [id=3, name=法官行为与涉诉信访研究, type=Type [id=4, bookType=法律与艺术]]
Book [id=4, name=巧用跨界思维学法律, type=Type [id=4, bookType=法律与艺术]]
Book [id=5, name=考研英语(二)词汇速记手册, type=Type [id=5, bookType=语言]]
```

图 10.13　多对一关联查询的运行结果

3. 一对多的关联

图书和图书类别是多对一的关系，反过来，就是图书类别对图书是一对多的关系，即一种图书类别对应多本图书。

实现表间一对多的关联查询的具体步骤如下：

①通过在 book 表中增加一个字段 type_id 作为外键，关联 type 表中的 id 字段，如图 10.11 和图 10.12 所示。

②创建对应的持久化对象 Type 和 Book，在 Type 对象中增加一个属性，关联 Book 对象，因为一个图书类型有多本图书，所以添加了一个 List 类型的对象，代码如下。

Type.java 文件：

```java
package pojo;
import java.util.List;
public class Type {
    private int id;
    private String bookType;
    private List<Book> books; //通过设置属性的方式与另一个对象建立关联
    public int getId() {
        return id;
    }
    public void setId(int id) {
        this.id = id;
    }
    public String getBookType() {
        return bookType;
    }
    public void setBookType(String bookType) {
        this.bookType = bookType;
    }
    public List<Book> getBooks() {
        return books;
    }
```

```java
    public void setBooks(List<Book> books) {
        this.books = books;
    }
    @Override
    public String toString() {
        return "Type [id=" + id + ", bookType=" + bookType + ", books=" + books + "]";
    }
}
```

Book.java 文件：

```java
package pojo;
public class Book {
    private int id;
    private String name;
    public int getId() {
        return id;
    }
    public void setId(int id) {
        this.id = id;
    }
    public String getName() {
        return name;
    }
    public void setName(String name) {
        this.name = name;
    }
    @Override
    public String toString() {
        return "Book [id=" + id + ", name=" + name + "]";
    }
}
```

③创建对 Type 对象操作的接口文件，创建相应的 SQL 映射文件。在 SQL 映射文件中添加 <select> 节点，编写相应的 SQL 语句，并构造关联映射结果。

接口文件 TypeDAO.java：

```java
package dao;
import java.util.List;
import pojo.Type;
public interface TypeDAO {
    List<Type> queryAllTypes();
}
```

SQL 映射文件 TypeDAO.xml：

```xml
<?xml version="1.0" encoding="UTF-8"?>
<!DOCTYPE mapper PUBLIC "-//mybatis.org//DTD Mapper 3.0//EN"
"http://mybatis.org/dtd/mybatis-3-mapper.dtd">
<mapper namespace="dao.TypeDAO">
    <select id="queryAllTypes" resultMap="TypeMap">
        select t.id tid,t.bookType type,b.id bid,b.name bookname from type t,book as b where b.type_id=t.id;
    </select>
    <resultMap type="pojo.Type" id="TypeMap">
        <id property="id" column="tid"/>
        <result property="bookType" column="type"/>
        <collection property="books" ofType="pojo.Book">
            <id property="id" column="bid"/>
            <result property="name" column="bookname"/>
        </collection>
    </resultMap>
</mapper>
```

④编写测试程序。

```java
import java.io.InputStream;
import java.util.List;
import javax.annotation.Resources;
import org.apache.ibatis.session.SqlSession;
import org.apache.ibatis.session.SqlSessionFactory;
import org.apache.ibatis.session.SqlSessionFactoryBuilder;
import dao.TypeDAO;
import pojo.Book;
import pojo.Type;
public class test {
    public static void main(String[] args) {
        //TODO Auto-generated method stub
        InputStream is = Resources.class.getResourceAsStream("/mybatis-config.xml");
        SqlSessionFactory sf = new SqlSessionFactoryBuilder().build(is);
        SqlSession sqlSession = sf.openSession();
        TypeDAO typeDao = sqlSession.getMapper(TypeDAO.class);
        List<Type> list = typeDao.queryAllTypes();
        for(Type t:list){
            System.out.println(t);
        }
    }
```

```
    }
}
```

⑤运行结果如图 10.14 所示。

```
Markers  Properties  Servers  Data Source Explorer  Snippets  Console
<terminated> test (3) [Java Application] C:\Program Files\Java\jdk-12.0.2\bin\javaw.exe (2021年6月11日 上午12:37:18 – 上午12:37:20)
Type [id=1, bookType=计算机与电子信息, books=[Book [id=1, name=java面试一战到底(基础卷)], Book [id=2, name=Java程序员面试笔试通关宝典]]]
Type [id=4, bookType=法律与艺术, books=[Book [id=3, name=法官行为与涉诉信访研究], Book [id=4, name=巧用跨界思维学法律]]]
Type [id=5, bookType=语言, books=[Book [id=5, name=考研英语(二)词汇速记手册]]]
```

图 10.14　一对多关联查询的运行结果

4. 多对多的关联

在学生选课系统中，一个学生可以选择多门课程，而一门课程可以被多个学生选择，所以学生和课程之间是多对多的关系。对于这种表间是多对多的关联查询，需要借助第三个表来建立关联，具体的步骤如下：

①创建三个数据表，即学生表 student、教师表 teacher 和选课表 sc，sc 作为中间表，将 student 表和 teacher 表关联起来，如图 10.15 至图 10.17 所示。

studentId	studentName
2010112101	王明
2010112102	李立
2010112103	林川

图 10.15　student 表

id	studentId	teacherId
1	2010112101	20030001
2	2010112101	20030045
3	2010112102	20030001
4	2010112102	20030012
5	2010112103	20030045
6	2010112103	20030012

图 10.16　sc 表

teacherId	teacherName
20030001	林老师
20030012	王老师
20030045	胡老师

图 10.17　teacher 表

②创建对应的持久化对象 Student、Sc 和 Teacher，在 Student 对象中增加一个属性，关联 Teacher 对象，因为一个学生可以选择多个老师，所以添加了一个 List < Teacher > 类型的属性，代码如下。

Student. java 文件：

```
package pojo;
import java.util.List;
public class Student {
    private String studentId;
    private String studentName;
    private List < Teacher > teachers; //通过设置属性的方式与另一个对象建立关联
    public String getStudentId() {
        return studentId;
    }
    public void setStudentId(String studentId) {
```

```java
        this.studentId = studentId;
    }
    public String getStudentName() {
        return studentName;
    }
    public void setStudentName(String studentName) {
        this.studentName = studentName;
    }
    public List<Teacher> getTeachers() {
        return teachers;
    }
    public void setTeachers(List<Teacher> teachers) {
        this.teachers = teachers;
    }
    @Override
    public String toString() {
        return "Student [studentId=" + studentId + ", studentName=" + studentName + ", teachers=" + teachers + "]";
    }
}
```

Teacher.java 文件:

```java
package pojo;
import java.util.List;
public class Teacher {
    private String teacherId;
    private String teacherName;
    private List<Student> students;  //通过设置属性的方式与另一个对象建立关联
    public String getTeacherId() {
        return teacherId;
    }
    public void setTeacherId(String teacherId) {
        this.teacherId = teacherId;
    }
    public String getTeacherName() {
        return teacherName;
    }
    public void setTeacherName(String teacherName) {
        this.teacherName = teacherName;
    }
    public List<Student> getStudents() {
```

```
        return students;
    }
    public void setStudents(List<Student> students){
        this.students = students;
    }
    @Override
    public String toString(){
        return "Teacher [teacherId = " + teacherId + ", teacherName = " + teacherName + "]";
    }
}
```

Sc.java 文件:

```
package pojo;
public class Sc{
    private Student student;
    private Teacher teacher;
    public Student getStudent(){
        return student;
    }
    public void setStudent(Student student){
        this.student = student;
    }
    public Teacher getTeacher(){
        return teacher;
    }
    public void setTeacher(Teacher teacher){
        this.teacher = teacher;
    }
    @Override
    public String toString(){
        return "Sc [student = " + student + ", teacher = " + teacher + "]";
    }
}
```

③创建对 Type 对象操作的接口文件,创建相应的 SQL 映射文件。在 SQL 映射文件中添加 <select> 节点,编写相应的 SQL 语句,并构造关联映射结果。

接口文件 StudentDAO.java:

```
package dao;
import java.util.List;
import pojo.Student;
public interface StudentDAO{
```

```
    List<Student> queryAllStudents();
}
```

SQL 映射文件 TypeDAO.xml：

```xml
<?xml version="1.0" encoding="UTF-8"?>
<!DOCTYPE mapper PUBLIC "-//mybatis.org//DTD Mapper 3.0//EN"
"http://mybatis.org/dtd/mybatis-3-mapper.dtd">
<mapper namespace="dao.StudentDAO">
    <select id="queryAllStudents" resultMap="StudentMap">
        select s.studentId,s.studentName,t.* from student s
            left outer join sc on s.studentId=sc.studentId
            left outer join teacher t on t.teacherId=sc.teacherId
    </select>
    <resultMap type="pojo.Student" id="StudentMap">
        <id property="studentId" column="studentId"/>
        <result property="studentName" column="studentName"/>
        <collection property="teachers" ofType="pojo.Teacher">
            <id property="teacherId" column="teacherId"/>
            <result property="teacherName" column="teacherName"/>
        </collection>
    </resultMap>
</mapper>
```

④编写测试程序。

```java
import java.io.InputStream;
import java.util.List;
import javax.annotation.Resources;
import org.apache.ibatis.session.SqlSession;
import org.apache.ibatis.session.SqlSessionFactory;
import org.apache.ibatis.session.SqlSessionFactoryBuilder;

import dao.StudentDAO;
import pojo.Student;
public class test {
    public static void main(String[] args) {
        //TODO Auto-generated method stub
        InputStream is = Resources.class.getResourceAsStream("/mybatis-config.xml");
        SqlSessionFactory sf = new SqlSessionFactoryBuilder().build(is);
        SqlSession sqlSession = sf.openSession();
        StudentDAO studentDao = sqlSession.getMapper(StudentDAO.class);
        List<Student> list = studentDao.queryAllStudents();
```

```
        for(Student s:list){
            System.out.println(s);
        }
    }
}
```

⑤运行结果如图 10.18 所示。

图 10.18　多对多的运行结果

10.3.3　动态 SQL

在编写 SQL 语句时,有的时候需要根据用户的操作实时拼接 SQL 语句,而不是使用固定不变的 select 语句,这就是动态 SQL。如在执行查询图书时,根据用户选择的查询条件是根据书名查询还是根据书号查询,从而确定查询条件,实时拼接成 SQL 语句。动态 SQL 是 Mybatis 框架的强大功能之一,在项目开发中,经常会用到这项功能。在讲解动态 SQL 时,为了能看到拼接后的 SQL 语句,需要利用日志来展示,下面先介绍如何设置日志。

1. 设置日志

先创建一个 Web 项目,命名为 mybatis_sql,整个项目的结构如图 10.19 所示。

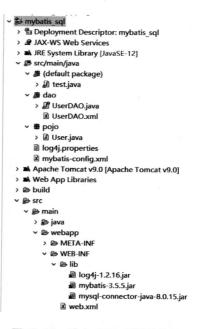

图 10.19　动态 SQL 项目结构图

Mybatis 框架默认使用 log4j 输出日志信息，在使用之前，要先对 log4j 进行配置。首先将 log4j-1.2.16.jar 包复制到 lib 目录中，然后在 src 目录下创建 log4j.properties 文件，对 log4j 进行配置，文件内容如图 10.20 所示。

```
# Global logging configuration
log4j.rootLogger=DEBUG, stdout
# Console output...
log4j.appender.stdout=org.apache.log4j.ConsoleAppender
log4j.appender.stdout.layout=org.apache.log4j.PatternLayout
log4j.appender.stdout.layout.ConversionPattern=%5p [%t] - %m%n
```

图 10.20　log4j.properties 文件内容

Log4j 包括三个重要的组成部分：日志信息的优先级、日志信息的输出目的地和日志信息的输出格式。日志信息的优先级从高到低分别为 FATAL、ERROR、WARN、INFO 和 DEBUG，可指定这条日志信息的重要程度；日志信息的输出目的地指日志输出到哪里，是输出到控制台、文件还是其他；输出格式则控制了日志信息的显示内容。

第一步：配置 rootLogger。

Log4j.rootLogger = [level], appenderName

level 是日志记录的优先级，一般使用 ERROR、WARN、INFO、DEBUG 四个级别，默认级别是 DEBUG。如果定义了 INFO 级别，则应用程序中所有 DEBUG 级别的日志信息将不被打印出来。appenderName 指定日志信息输出目的地的名称。

上面配置文件中设置为 Log4j.rootLogger = DEBUG, stdout 表示输出 DEBUG 级别的日志信息，输出目的地名称是 stdout。

第二步：配置日志信息输出目的地。

Log4j 提供以下几种输出目的地：

org.apache.log4j.ConsoleAppender　　控制台

org.apache.log4j.FileAppender　　文件

org.apache.log4j.DaiyRollingFileAppender　　每天产生一个日志文件

org.apache.log4j.RollingFileAppender　　文件达到指定大小的时候产生一个新的文件

org.apache.log4j.WriterAppender　　将日志信息以流格式发送到任意指定的地方

上面配置文件设置了 log4j.appender.stdout = org.apache.log4j.ConsoleAppender，输出目的地 stdout 表示的是控制台。

第三步：配置日志信息的格式，即布局。

Log4j 提供以下几种布局：

org.apache.log4j.HTMLLayout　　以 HTML 表格形式布局

org.apache.log4j.PatternLayout　　可以灵活地指定布局模式

org.apache.log4j.SimpleLayout　　包含日志信息的级别和信息字符串

org.apache.log4j.TTCCLayout　　包含日志产生的时间、线程和类别等信息

Log4J 采用类似 C 语言中的 printf 函数的打印格式来格式化日志信息。

%p:输出优先级,即DEBUG、INFO、WARN、ERROR和FATAL。

%r:输出自应用启动到输出该log信息所耗费的毫秒数。

%c:输出所属的类目,通常就是所在类的全名。

%t:输出产生该日志事件的线程名。

%n:输出一个回车换行符,Windows平台为"rn",Unix平台为"n"。

%d:输出日志时间点的日期或时间,默认格式为ISO 8601,也可以在其后指定格式,比如%d{yyyy mmm dd HH:mm:ss,SSS},输出类似2021年3月25日10:07,921。

%l:输出日志事件的发生位置,包括类目名、发生的线程及代码中的行,例如Testlog4.main(TestLog4.java:10)。

上面配置文件设置了log4j.appender.stdout.layout = org.apache.log4j.PatternLayout,表示可以灵活地指定布局模式。

```
log4j.appender.stdout.layout.ConversionPattern=%5p [%t]-%m%n
```

其中,%5p表示输出该日志的优先级,%t表示输出产生该日志事件的线程名,%m表示输出代码中指定的消息,%n表示输出一个回车换行符。

2. <if>和<where>元素

为了介绍方便,还是继续使用user表,查询用户时可以根据id值或username值或password值或它们之间的组合来实现。

(1)单独根据id值查询,SQL语句是select * from user where id = #{id}。

(2)单独根据username值查询,SQL语句是select * from user where username = #{username}。

(3)单独根据password值查询,SQL语句是select * from user where password = #{password}。

(4)组合条件查询可以是select * from user where id = #{id} and username = #{username}或select * from user where id = #{id} and password = #{password}或select * from user where username = #{username} and password = #{password}或select * from user 等。

在项目运行时,根据用户填入的条件实时地拼接SQL语句时需要使用<if>元素进行判断,并使用<where>元素构造条件。

接口文件UserDAO.java:

```
package dao;
import java.util.List;
import org.apache.ibatis.annotations.Param;
import pojo.User;
public interface UserDAO {
    public List<User> query(@Param("id")int id,@Param("username")String username,@Param("password")String password);
}
```

在 SQL 映射文件 UserDAO.xml 中，select 语句书写如下：

```xml
<select id="query" resultType="pojo.User">
   select * from user
   <where>
      <if test="username!=null and username!=''">
         and username = #{username}
      </if>
      <if test="password!=null and password!=''">
         or password = #{password}
      </if>
      <if test="id!=0">
         and id = #{id}
      </if>
   </where>
</select>
```

编写测试文件 test.java：

```java
public class test {
   public static void main(String[] args) {
       InputStream is = Resources.class.getResourceAsStream("/mybatis-config.xml");
       SqlSessionFactory sf = new SqlSessionFactoryBuilder().build(is);
       SqlSession sqlSession = sf.openSession();
       UserDAO userDao = sqlSession.getMapper(UserDAO.class);
       List<User> list = userDao.query(0,"a","e");
       for(User u:list){
          System.out.println(u);
       }
   }
}
```

数据库 user 表如图 10.21 所示。

id	username	password
1	a	a
2	b	b
3	c	c
4	d	d
5	e	e

图 10.21 user 表

运行后打开控制台，可以看到输出的日志如图 10.22 所示。

```
<terminated> test (6) [Java Application] C:\Program Files\Java\jdk-12.0.2\bin\javaw.exe (2021年6月11日 下午2:58:36 – 下午2:58:37)
=======query======
DEBUG [main] - Logging initialized using 'class org.apache.ibatis.logging.log4j.Log4jImpl' adapter.
DEBUG [main] - PooledDataSource forcefully closed/removed all connections.
DEBUG [main] - PooledDataSource forcefully closed/removed all connections.
DEBUG [main] - PooledDataSource forcefully closed/removed all connections.
DEBUG [main] - PooledDataSource forcefully closed/removed all connections.
DEBUG [main] - Opening JDBC Connection
DEBUG [main] - Created connection 844112759.
DEBUG [main] - Setting autocommit to false on JDBC Connection [com.mysql.cj.jdbc.ConnectionImpl@32502377]
DEBUG [main] - ==>  Preparing: select * from user WHERE username=? or password=?
DEBUG [main] - ==> Parameters: a(String), e(String)
DEBUG [main] - <==      Total: 1
User [id=1, username=a, password=a]
```

图 10.22　使用 <if> 和 <where> 元素实现动态 SQL 控制台输出的日志

通过控制台输出的日志可以看到，当 username 和 password 都不为 0 时，Mybatis 框架会实时拼接 SQL 语句，拼接成 select * from user where and username = ? or password = ?，<where>元素会自动地将多出的 and 删除，就变成了日志中输出的 SQL 语句"select * from user where username = ? or password = ?"。

3. <foreach>元素

在很多项目中，对数据表的删除操作都实现了批量删除，也就是同时删除多条记录，通过将前端用户选择的多条记录保存为数组，传递到服务器端交给 Mybatis 框架处理，Mybatis 框架利用<foreach>元素，拼接 SQL 语句，具体代码如下：

接口文件 UserDAO.java：

```
package dao;
import java.util.List;
import org.apache.ibatis.annotations.Param;
import pojo.User;
public interface UserDAO {
    public int deleteByIds(@ Param("ids")List ids);
}
```

SQL 映射文件 UserDAO.xml：

```
<?xml version = "1.0" encoding = "UTF-8"? >
<!DOCTYPE mapper
    PUBLIC " - //mybatis.org//DTD Mapper 3.0//EN"
    "http://mybatis.org/dtd/mybatis-3-mapper.dtd" >
<mapper namespace = "dao.UserDAO" >
    <! - - delete from user where id in(1,2,5) - - >
    <delete id = "deleteByIds" >
    delete from user where id in
    <foreach collection = "ids" item = "id" open = "(" separator = "," close = ")" >
        #{id}
    </foreach>
```

```
    </delete>
</mapper>
```

测试文件如下：

```
public class test {
    public static void main(String[] args) {
InputStream is = Resources.class.getResourceAsStream("/mybatis-config.xml");
        SqlSessionFactory sf = new SqlSessionFactoryBuilder().build(is);
        SqlSession session = sf.openSession();
        UserDAO userDAO = session.getMapper(UserDAO.class);
        List ids = new ArrayList();
        ids.add(1);
        ids.add(2);
        ids.add(4);
        int i = userDAO.deleteByIds(ids);
        session.commit();
        session.close();
    }
}
```

运行后控制台的输出如图 10.23 所示，可以看到拼接后的 SQL 语句为"delete from user where id in (?,?,?)"。

```
<terminated> test (6) [Java Application] C:\Program Files\Java\jdk-12.0.2\bin\javaw.exe (2021年6月11日 下午3:01:21 – 下午3:01:22)
=======deleteByIds======
DEBUG [main] - Logging initialized using 'class org.apache.ibatis.logging.log4j.Log4jImpl' adapter.
DEBUG [main] - PooledDataSource forcefully closed/removed all connections.
DEBUG [main] - PooledDataSource forcefully closed/removed all connections.
DEBUG [main] - PooledDataSource forcefully closed/removed all connections.
DEBUG [main] - PooledDataSource forcefully closed/removed all connections.
DEBUG [main] - Opening JDBC Connection
DEBUG [main] - Created connection 1213818572.
DEBUG [main] - Setting autocommit to false on JDBC Connection [com.mysql.cj.jdbc.ConnectionImpl@485966cc]
DEBUG [main] - ==>  Preparing: delete from user where id in ( ? , ? , ? )
DEBUG [main] - ==> Parameters: 1(Integer), 2(Integer), 4(Integer)
DEBUG [main] - <==    Updates: 2
DEBUG [main] - Committing JDBC Connection [com.mysql.cj.jdbc.ConnectionImpl@485966cc]
DEBUG [main] - Resetting autocommit to true on JDBC Connection [com.mysql.cj.jdbc.ConnectionImpl@485966cc]
DEBUG [main] - Closing JDBC Connection [com.mysql.cj.jdbc.ConnectionImpl@485966cc]
DEBUG [main] - Returned connection 1213818572 to pool.
```

图 10.23 使用 <foreach> 元素实现动态 SQL 控制台输出的日志

从图 10.24 可以看到，已删除指定的 id 为 1、2、4 三条记录。

id	username	password
3	c	c
5	e	e

图 10.24 删除指定 id 号后的 user 表

<foreach>元素也可以用在select语句中,用于查询多条记录。

4. <set>元素

利用<set>元素可以实时拼接update语句,当字段有修改就进行更新,没有修改就不进行更新。

接口文件UserDAO.java:

```java
package dao;
import java.util.List;
import org.apache.ibatis.annotations.Param;
import pojo.User;
public interface UserDAO {
    public int update(@Param("username")String username,@Param("password")String password,@Param("id")int i);
}
```

SQL映射文件UserDAO.xml:

```xml
<?xml version="1.0" encoding="UTF-8"?>
<!DOCTYPE mapper
    PUBLIC "-//mybatis.org//DTD Mapper 3.0//EN"
    "http://mybatis.org/dtd/mybatis-3-mapper.dtd">
<mapper namespace="dao.UserDAO">
    <update id="update">
        update user
        <set>
            <if test="username!=null and username!=''">
                username=#{username}
            </if>
            <if test="password!=null and password!=''">
                password=#{password}
            </if>
            where id=#{id};
        </set>
    </update>
</mapper>
```

测试文件如下:

```java
public class test {
    public static void main(String[] args) {
        InputStream is = Resources.class.getResourceAsStream("/mybatis-config.xml");
        SqlSessionFactory sf = new SqlSessionFactoryBuilder().build(is);
        SqlSession session = sf.openSession();
        UserDAO userDAO = session.getMapper(UserDAO.class);
        int i = userDAO.update("ee","",5);
```

```
        session.commit();
        session.close();
    }
}
```

运行后控制台输出日志如图 10.25 所示。

```
<terminated> test (6) [Java Application] C:\Program Files\Java\jdk-12.0.2\bin\javaw.exe (2021年6月11日 下午3:09:24 – 下午3:09:26)
=============update========
DEBUG [main] - Logging initialized using 'class org.apache.ibatis.logging.log4j.Log4jImpl' adapter.
DEBUG [main] - PooledDataSource forcefully closed/removed all connections.
DEBUG [main] - PooledDataSource forcefully closed/removed all connections.
DEBUG [main] - PooledDataSource forcefully closed/removed all connections.
DEBUG [main] - PooledDataSource forcefully closed/removed all connections.
DEBUG [main] - Opening JDBC Connection
DEBUG [main] - Created connection 424398527.
DEBUG [main] - Setting autocommit to false on JDBC Connection [com.mysql.cj.jdbc.ConnectionImpl@194bcebf]
DEBUG [main] - ==>  Preparing: update user SET username=? where id=?;
DEBUG [main] - ==> Parameters: ee(String), 5(Integer)
DEBUG [main] - <==    Updates: 0
DEBUG [main] - Committing JDBC Connection [com.mysql.cj.jdbc.ConnectionImpl@194bcebf]
DEBUG [main] - Resetting autocommit to true on JDBC Connection [com.mysql.cj.jdbc.ConnectionImpl@194bcebf]
DEBUG [main] - Closing JDBC Connection [com.mysql.cj.jdbc.ConnectionImpl@194bcebf]
DEBUG [main] - Returned connection 424398527 to pool.
```

图 10.25 使用 <set> 元素实现动态 SQL 控制台输出的日志

可以看到，在测试文件中，若只传入 username 的参数，而 password 参数为空，则动态拼接 SQL 语句时，只设置了 username = ?。

习　　题

使用 Mybatis 框架实现数据库的增删改查操作，数据库名为 bookmanage，表为 user，包含 id、username、password 字段。

第 11 章　SSM 框架整合

【本章内容】

- 11.1　整合的步骤
- 11.2　图书管理系统功能描述
- 11.3　各层的规划与参数传递
- 11.4　部分功能的实现

在这一章,我们将对前面介绍的几个框架进行整合,从整合的步骤、相关的配置到选用图书管理系统的开发作为案例进行介绍。在图书管理系统案例中,着重介绍控制器、业务逻辑层、数据访问层之间参数的传递和文件的命名规范,以此引导大家使用 SSM 进行项目开发。

11.1　整合的步骤

下面介绍 Spring、SpringMVC 和 Mybatis 三个框架整合的步骤。因为 SpringMVC 是 Spring 框架的一个模块,所以没有所谓的整合,则只需要时 Spring 框架和 Mybatis 框架进行整合,这需要额外增加一个包 mybatis-spring-1.3.1.jar。

11.1.1　创建项目和相关的包、文件夹

创建一个 Dynamic Web Project 类型的项目,命名为 bookmanage,在 src 目录中创建 com.books.controller、com.books.service、com.books.dao 和 com.books.pojo 包,再创建一个 config 文件夹,如图 11.1 所示。

11.1.2　导入相关的包

在 webapp 目录下的 lib 目录中导入相关的包,下面列出的是基本的包,实际开发时根据项目的需要,还要再导入其他包。

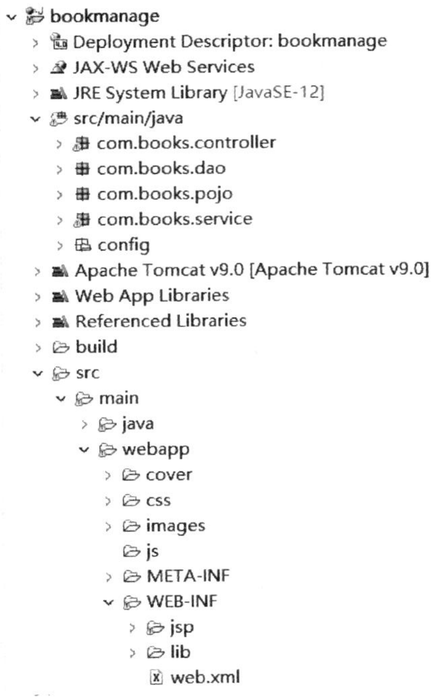

```
> bookmanage
  > Deployment Descriptor: bookmanage
  > JAX-WS Web Services
  > JRE System Library [JavaSE-12]
  v src/main/java
    > com.books.controller
    > com.books.dao
    > com.books.pojo
    > com.books.service
    > config
  > Apache Tomcat v9.0 [Apache Tomcat v9.0]
  > Web App Libraries
  > Referenced Libraries
  > build
  v src
    v main
      > java
      v webapp
        > cover
        > css
        > images
          js
        > META-INF
        v WEB-INF
          > jsp
          > lib
            web.xml
```

图 11.1　项目目录结构

1. Spring 框架和 SpringMVC 框架 jar 包

Spring 框架 4 个核心包：

spring-beans-5.3.6.jar

spring-context--5.3.6.jar

spring-core-5.3.6.jar

spring-expression--5.3.6.jar

SpringMVC 框架包：

spring-web-5.3.6.jar

spring-webmvc-5.3.6.jar

aop 开发包：

aopalliance-1.0.jar

aspectweave-1.8.10.jar

spring-aop-5.3.6.jar

spring-aspects-5.3.6.jar

JDBC 包：

spring-jdbc-5.3.6.jar

事务管理包：

spring-tx-5.3.6.jar

上传文件包：

commons-fileupload-1.3.2.jar

commons-io-2.5.jar

2. Mybatis 框架 jar 包和数据库驱动程序 jar 包

Mybatis 框架包：

Mybatis-3.5.5.jar

MySQL 数据库驱动包：

Mysql-connector-Java-8.0.15.jar

Mybatis 框架和 Spring 框架整合包：

Mybatis-spring-1.3.1.jar

3. Mybatis 的依赖包

commons-logging-1.2.jar

ant-1.9.6.jar

ant-launcher-1.9.6.jar

asm-5.1.jar

cglib-3.2.4.jar

javassist-3.21.0-GA.jar

log4j-1.2.17.jar

log4j-api-2.3.jar

log4j-core-2.3.jar

ognl-3.1.12.jar

slf4j-api-1.7.22.jar

slf4j-log4j12-1.7.22.jar

4. 数据源所需 jar 包

commons-dbcp2-2.1.1.jar

commons-pool2-2.4.2.jar

5. JSTL 标准标签库的 jar 包

jstl-1.2.jar

11.1.3　创建配置文件

在 config 文件夹中，创建数据库常量配置文件 db.properties、Spring 框架的配置文件 applicationContext.xml 和 SpringMVC 框架的配置文件 springmvc-config.xml，并编辑项目配置文件 web.xml。

1. 项目配置文件 web.xml

```
<?xml version = "1.0" encoding = "UTF-8"? >
<web-app xmlns = "http://xmlns.jcp.org/xml/ns/Javaee"
    xmlns:xsi = "http://www.w3.org/2001/XMLSchema-instance"
    xsi:schemaLocation = "http://xmlns.jcp.org/xml/ns/Javaee http://xmlns.jcp.org/xml/ns/Javaee/web-app_4_0.xsd"
```

```xml
    version = "4.0" >
<display-name>bookmanage</display-name>
<welcome-file-list>
<welcome-file>index.jsp</welcome-file>
</welcome-file-list>
<!-- 配置前端控制器,并通过参数指明配置文件 springmvc-config.xml 的位置 -->
<servlet>
    <servlet-name>springmvc</servlet-name>    <servlet-class>org.springframework.web.servlet.DispatcherServlet</servlet-class>
    <init-param>
    <param-name>contextConfigLocation</param-name>
        <param-value>classpath:config/springmvc-config.xml</param-value>
    </init-param>
</servlet>
<servlet-mapping>
    <servlet-name>springmvc</servlet-name>
    <url-pattern>/</url-pattern>
</servlet-mapping>
<!-- 配置监听器 ContextLoaderListener 来加载 Spring 框架的配置文件 applicationContext.xml -->
<context-param>
        <param-name>contextConfigLocation</param-name>    <param-value>classpath:config/applicationContext.xml</param-value>
</context-param>
<listener>
        <listener-class>org.springframework.web.context.ContextLoaderListener</listener-class>
    </listener>
</web-app>
```

在默认的 web.xml 文件中增加了前端控制器 DispatcherServletr 的配置和监听器 ContextLoaderListener 的配置。

2. 数据库常量配置文件 db.properties

```
jdbc.driver = com.mysql.cj.jdbc.Driver
jdbc.url = jdbc:mysql://localhost:3306/bookmanage? useUnicode = true&characterEncoding = UTF-8&serverTimezone = Asia/Shanghai
jdbc.username = root
jdbc.password = 123456
```

在这个文件中,设置了 JDBC 驱动器的名称、连接的 URL 地址、用户名和密码,URL 地址后面带 useUnicode 和 characterEncoding 两个参数是为了解决中文乱码的问题。这些信息独立在这个文件的好处在于当这些信息要修改时,直接在文件里修改就可以,比较方便。

3. Spring框架配置文件applicationContext.xml

```xml
<?xml version="1.0" encoding="UTF-8"?>
<beans xmlns="http://www.springframework.org/schema/beans"
    xmlns:xsi="http://www.w3.org/2001/XMLSchema-instance"
    xmlns:aop="http://www.springframework.org/schema/aop"
    xmlns:tx="http://www.springframework.org/schema/tx"
    xmlns:context="http://www.springframework.org/schema/context"
    xsi:schemaLocation="http://www.springframework.org/schema/beans
    http://www.springframework.org/schema/beans/spring-beans-4.3.xsd
    http://www.springframework.org/schema/tx
    http://www.springframework.org/schema/tx/spring-tx-4.3.xsd
    http://www.springframework.org/schema/context
    http://www.springframework.org/schema/context/spring-context-4.3.xsd
    http://www.springframework.org/schema/aop
    http://www.springframework.org/schema/aop/spring-aop-4.3.xsd">
<!--读取数据库常量配置文件-->
<context:property-placeholder location="classpath:config/db.properties"/>
<!--配置数据源-->
<bean id="dataSource" class="org.springframework.jdbc.datasource.DriverManagerDataSource">
    <property name="driverClassName" value="${jdbc.driver}"/>
    <property name="url" value="${jdbc.url}"/>
    <property name="username" value="${jdbc.username}"/>
    <property name="password" value="${jdbc.password}"/>
</bean>
<!--配置SqlSessionFactory-->
<bean id="sqlSessionFactory" class="org.mybatis.spring.SqlSessionFactoryBean">
    <property name="dataSource" ref="dataSource"/>
    <property name="mapperLocations" value="classpath:com/books/dao/*Dao.xml"/>
    <property name="typeAliasesPackage" value="com/books/pojo"/>
</bean>
<!--配置mapper扫描器-->
<bean class="org.mybatis.spring.mapper.MapperScannerConfigurer">
    <property name="basePackage" value="com.books.dao"/>
</bean>
<!--扫描service-->
<context:component-scan base-package="com.books.service"/>
</beans>
```

首先配置了读取数据库常量配置文件 db.properties，接着配置数据源，所需要的连接参数通过 db.properties 读取，最后，再配置 SqlSessionFactory，指明其依赖的数据源，并指明映射文件。

对于 SSM 框架整合，在 applicationContext.xml 文件中已配置了数据源和映射文件，所以不再需要创建 Mybatis 框架的配置文件 mybatis-config.xml。这里让我们再回顾一下 mybatis-config.xml 的内容。

```xml
<?xml version="1.0" encoding="UTF-8"?>
<!DOCTYPE configuration PUBLIC "-//mybatis.org//DTD Config 3.0//EN"
"http://mybatis.org/dtd/mybatis-3-config.dtd">
<configuration>
    <environments default="mysql">
        <environment id="mysql">
            <transactionManager type="JDBC"/>
            <dataSource type="POOLED">
                <property name="driver" value="com.mysql.cj.jdbc.Driver"/>
                <property name="url" value="jdbc:mysql://localhost:3306/jdbctest?useUnicode=true&characterEncoding=UTF-8&serverTimezone=Asia/Shanghai"/>
                <property name="username" value="root"/>
                <property name="password" value="123456"/>
            </dataSource>
        </environment>
    </environments>
    <mappers>
        <mapper resource="com/books/dao/UserDAO.xml"/>
    </mappers>
</configuration>
```

可以看到，mybatis-config.xml 文件就是用于配置数据库连接信息和设置 SQL 映射文件。

4. SpringMVC 框架配置文件 springmvc-config.xml

```xml
<?xml version="1.0" encoding="UTF-8"?>
<beans xmlns="http://www.springframework.org/schema/beans"
    xmlns:xsi="http://www.w3.org/2001/XMLSchema-instance"
    xmlns:context="http://www.springframework.org/schema/context"
    xmlns:mvc="http://www.springframework.org/schema/mvc"
    xsi:schemaLocation="http://www.springframework.org/schema/beans http://www.springframework.org/schema/beans/spring-beans-4.3.xsd
    http://www.springframework.org/schema/context http://www.springframework.org/schema/context/spring-context-4.3.xsd
    http://www.springframework.org/schema/mvc http://www.springframework.org/schema/mvc/spring-mvc-4.3.xsd">
<!-- 配置注解驱动 -->
```

```xml
<mvc:annotation-driven/>
<!-- 配置包扫描器,扫描@Controller注解的类 -->
<context:component-scan base-package="com.books.controller"/>
<!-- 配置视图解析器 -->
<bean class="org.springframework.web.servlet.view.InternalResourceViewResolver">
    <property name="suffix" value=".jsp"/>
    <property name="prefix" value="/WEB-INF/jsp/"/>
</bean>
<!-- 配置静态资源的访问映射,此配置中的文件,将不被前端控制器拦截 -->
<mvc:resources location="/cover/" mapping="/cover/**"/>
<mvc:resources location="/images/" mapping="/images/**"/>
<mvc:resources location="/css/" mapping="/css/**"/>
<!-- 定义拦截器 -->
<mvc:interceptors>
    <mvc:interceptor>
        <mvc:mapping path="/*"/>
        <!-- 使用bean定义一个Interceptor,直接定义在mvc:interceptors根下面的Interceptor将拦截所有的请求 -->
        <bean class="com.books.controller.AuthorizationInterceptor"/>
    </mvc:interceptor>
</mvc:interceptors>
<!-- 配置文件上传解析器MultipartResolver -->
<bean id="multipartResolver" class="org.springframework.web.multipart.commons.CommonsMultipartResolver">
    <!-- 设置编码格式 -->
    <property name="defaultEncoding" value="UTF-8"/>
</bean>
</beans>
```

11.1.4 配置静态资源访问方式

在web.xml文件中,前端控制器配置的"/"会将页面中引入的静态资源进行拦截,这些静态资源包括js文件、css文件,还有网页所依赖的一些图片、音频和视频文件,而拦截后页面将找不到这些静态资源文件,会引起页面报错。有以下三种方式可以访问静态资源。

1. `<mvc:resources location="/js/" mapping="/js/**"/>`

在springmvc-config.xml配置文件中,增加`<mvc:resources>`标签,配置静态资源的访问映射,配置所指定的文件,将不被前端控制器拦截。

2. 使用`<mvc:default-servlet-handler>`标签

在springmvc-config.xml文件中,使用`<mvc:default-servlet-handler>`标签后,会在SpringMVC 上下文中定义一个 org. springframework. web. servlet. resource. DefaultHttp-Request-Handler（即默认的 Servlet 请求处理器），对进入 DispatcherServlet 的 URL 进行筛查，如果发现是静态资源的请求，就将该请求转由 Web 服务器默认的 Servlet 处理，默认的 Servlet 就会对这些静态资源放行。

3. 激活 Tomcat 默认的 Servlet 来处理静态资源访问

需要在 web. xml 文件中添加如下内容。

```
<servlet-mapping>
    <servlet-name>default</servlet-name>
    <url-pattern>*.js</url-pattern>
</servlet-mapping>
<servlet-mapping>
    <servlet-name>default</servlet-name>
    <url-pattern>*.css</url-pattern>
</servlet-mapping>
...
```

最简单的方式就是第 2 种方式，直接一行代码就可以把所有路径下所有类型的静态资源放行，不再拦截，而第 1 种方式也用得比较多，但需要具体地指定哪个路径中的哪些静态文件放行。

11.1.5 配置编码过滤器

前端页面提交的表单数据如果是中文字符，这时要通过 request. setCharacterEncoding("utf-8")将编码设置成能够支持中文字符。在项目开发中，需要多次进行设置，比较麻烦。Spring 框架提供了编码过滤器，只需在 web. xml 文件中进行配置，就可以对整个页面进行统一处理，非常方便。

```
<!-- 配置编码过滤器 -->
<filter>
    <filter-name>CharacterEncodingFilter</filter-name>
    <filter-class>org.springframework.web.filter.CharacterEncodingFilter</filter-class>
    <init-param>
        <param-name>encoding</param-name>
        <param-value>UTF-8</param-value>
    </init-param>
</filter>
<filter-mapping>
    <filter-name>CharacterEncodingFilter</filter-name>
    <url-pattern>/*</url-pattern>
</filter-mapping>
```

Spring 框架提供的过滤器是 org. SpringFramework. web. filter. CharacterEncodingFilter，对

所有页面的请求信息都进行编码设置。

11.2 图书管理系统功能描述

11.2.1 功能介绍

下面利用 SSM 框架开发图书管理系统,项目命名为 bookmanage。把前面介绍的知识综合运用,此图书管理系统具有游客和管理员两种用户。

游客:具有浏览图书和根据书名、类别查询图书的功能。

管理员:管理员必须要登录验证后才具有增删改的功能,可添加新图书,并上传图书封面。

11.2.2 数据库设计

数据库命名为 bookmanage,表 book 包括 isbn,title,author,type_id,publisher,price,publicationDate,introduction,photo 字段;表 type 包括 id,typename 字段;表 user 包括 id,username,password 字段。三个表的结构如表 11.1、表 11.2 和表 11.3 所示。

表 11.1 book 表(图书信息表)

字段名	类型	长度	小数点	是否主键	说明
isbn	varchar	100	0	是	书号
title	varchar	100	0	否	书名
type_id	int	11	0	否	类型 id
author	varchar	100	0	否	作者
introduction	varchar	255	0	否	简介
price	double	10	2	否	单价
publisher	varchar	100	0	否	出版社
photo	varchar	100	0	否	封面
publishcationDate	date		0	否	出版日期

表 11.2 type 表(类型表)

字段名	类型	长度	小数点	是否主键	说明
id	int	11	0	是	类型编号
typename	varchar	100	0	否	类型名

表 11.3　user 表（用户表）

字段名	类型	长度	小数点	是否主键	说明
id	int	11	0	是	用户编号
username	varchar	100	0	否	用户名
password	varchar	100	0	否	密码

11.3　各层规划与参数传递

11.3.1　各层规划

在进行项目开发时，要先做好系统功能所需要的各层文件和类方法的命名。图 11.2 展示了系统设计的规划。

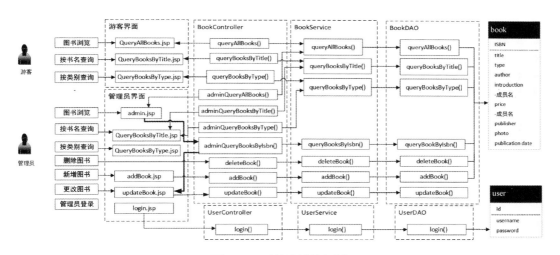

图 11.2　系统设计的规划

系统设计规划里面的箭头表示方向，可以看到很多功能的实现是从视图→控制器→业务逻辑层→数据访问层，完成之后再将数据传递给视图。而更新功能需要分两步来实现，先按 ISBN 号查询，然后回到更改视图，更改好之后，交给实现更改的控制器，再一步步调用来实现。

11.3.2　层与层之间参数的传递

在系统设计规划图中，由于篇幅的原因，没能够把各层方法中的参数都表示出来，在编码实现时，这些参数的传递非常重要，这一节将各个功能实现在各层之间的参数传递表示出来，便于大家对前面 SpringMVC 框架和 Mybatis 框架各部分内容加深理解。

1. 管理员登录功能(图 11.3)

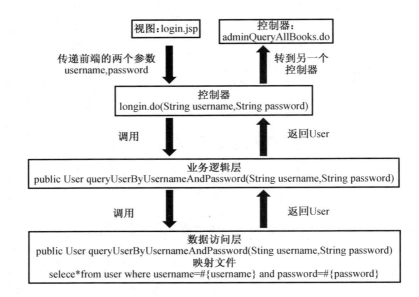

图 11.3 登录功能

2. 游客浏览图书功能(图 11.4)

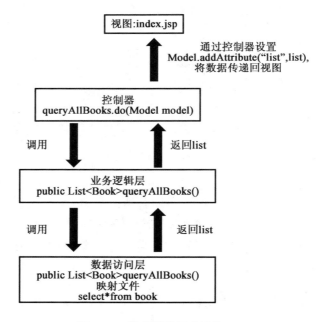

图 11.4 游客浏览图书功能

3. 游客或管理员按书名搜索图书功能(图 11.5)

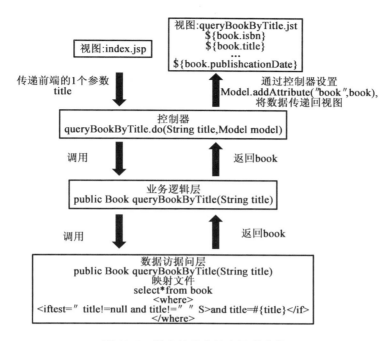

图 11.5 游客按书名搜索图书功能

4. 游客或管理员按类别搜索图书功能(图 11.6)

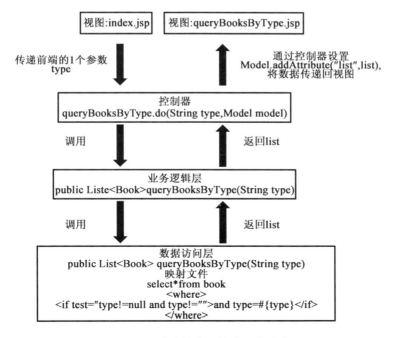

图 11.6 游客按类别搜索图书功能

5. 管理员添加图书功能（图 11.7）

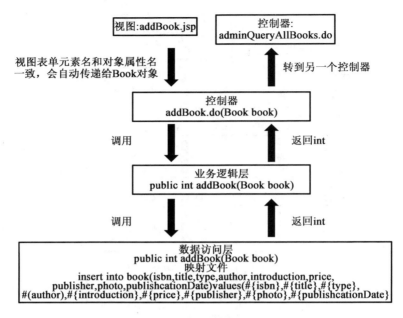

图 11.7　管理员添加图书功能

6. 管理员删除图书功能（图 11.8）

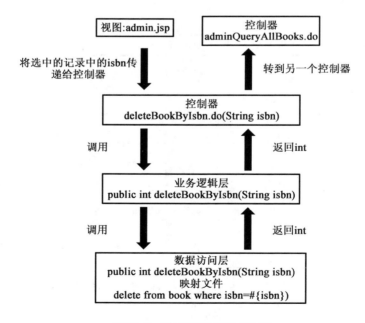

图 11.8　管理员删除图书功能

7. 管理员更改图书功能

第一步:按 ISBN 号搜索图书功能,如图 11.9 所示。

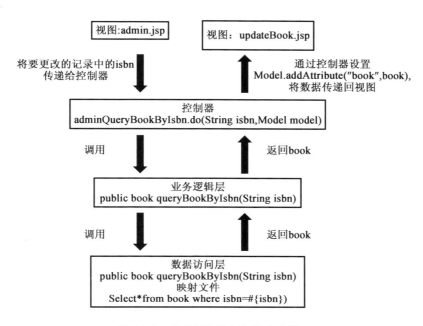

图 11.9　按 ISBN 号查询图书功能

第二步:对指定 ISBN 号的图书信息进行更改,如图 11.10 所示。

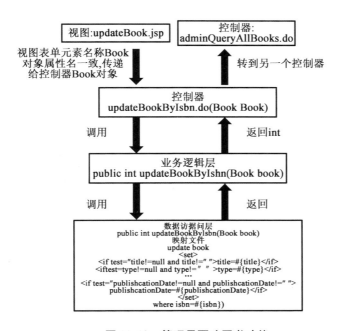

图 11.10　管理员更改图书功能

11.4 部分功能的实现

11.4.1 管理主页面的实现

管理员登录后进入的页面为管理主页面,页面显示了所有图书,并增加了对图书的删除、编辑、查询等操作。用户操作的流程是先登录,然后执行查询所有图书的控制器,再转到主页面视图文件 admin.jsp。现在按照上一节的"各层规划",由底往上逐层介绍。

1. 持久化类

在 pojo 包创建持久化类 Book.java,里面定义的属性与数据表 book 的字段是对应的,属性名和字段名相同,属性的类型和字段的类型是对应的,这里省略代码。

2. 数据访问层

在 dao 包中创建数据访问层接口类 BookDAO.java,然后再创建相应的映射文件 BookDAO.xml,在 BookDAO.java 文件中定义实现管理页面查询所有图书的方法,并在映射文件中编写相应的 SQL 语句。

BookDAO.java 文件:

```
package com.books.dao;
import com.books.pojo.Book;
import java.util.List;
import org.apache.ibatis.annotations.Param;
public interface BookDAO{
    public List<Book> queryAllBooks(@Param("pageNo") Integer pageNo, @Param("pageSize") Integer pageSize);
}
```

QueryAllBooks()方法带了两个参数,用于实现分页,如果不实现分页,查询所有图书的方法是没有参数的。

BookDAO.xml 文件:

```
<?xml version="1.0" encoding="UTF-8"?>
<!DOCTYPE mapper PUBLIC "-//mybatis.org//DTD Mapper 3.0//EN" "http://mybatis.org/dtd/mybatis-3-mapper.dtd">
<mapper namespace="com.books.dao.BookDao">
    <select id="queryAllBooks" resultType="com.books.pojo.Book">
      select * from book  order by publicationDate limit #{pageNo},#{pageSize}
    </select>
    </select>
</mapper>
```

在映射文件中,查询使用的是 select 语句,后面加上"limit #{pageNo},#{pageSize}"实现分页的效果,#{pageNo}参数设置了从哪页开始,#{pageSize}参数设置了每页有多少条记录。实现分页有三种方式，本书采用的是最简单的方式；第二种方法是利用 interceptor 拼接 SQL，实现和 limit 一样的功能；第三种方式就是利用 Mybatis 框架的 PageHelper 插件实现分页，不同版本的 pagehelper 配置上不太相同，需要的话，可以上网查找相关资料。

3. 业务逻辑层

业务逻辑层通过调用数据访问层实现业务。在 service 包创建图书管理的业务逻辑层接口文件 BookService.java，再创建实现类文件 BookServiceImpl.java。

BookService.java 文件：

```java
package com.books.service;
import com.books.pojo.Book;
import java.util.List;
public interface BookService {
    public List<Book> adminQueryAllBooks();
}
```

BookServiceImpl.java 文件：

```java
package com.books.service;
import com.books.dao.BookDAO;
import com.books.pojo.Book;
import java.util.List;
import org.springframework.beans.factory.annotation.Autowired;
import org.springframework.stereotype.Service;
@Service
public class BookServiceImpl implements BookService {
    @Autowired
    private BookDAO bookDao;
    public List<Book> queryAllBooks(int pageNo, int pageSize){
        return this.bookDao.queryAllBooks(pageNo,pageSize);
    }
}
```

BookServiceImpl 是接口 BookService 的实现类，@Service 注解是一个业务逻辑层组件，在实现类中，业务逻辑层需要调用数据访问层，各层的组件都由 Spring 容器管理，控制反转。数据访问层组件 BookDAO 作为业务逻辑层的依赖，通过属性的方式注入，使用了 @Autowired 表示自动装配的注入方式。

4. 控制层

在 controller 包中创建控制器 BookController.java，实现业务操作，代码如下。

```java
package com.books.controller;
import com.books.pojo.Book;
import com.books.service.BookService;
import com.books.service.UserService;
import java.util.List;
```

```java
import org.springframework.beans.factory.annotation.Autowired;
import org.springframework.stereotype.Controller;
import org.springframework.ui.Model;
import org.springframework.web.bind.annotation.RequestMapping;
@Controller
public class BookController {
    @Autowired
    public BookService bookService;
    @Autowired
    private UserService userService;
    @RequestMapping({"/queryAllBooks"})
    public String queryAllBooks(int pageNo, int pageSize, Model model) {
        List<Book> books = this.bookService.queryAllBooks(pageNo, pageSize);
        model.addAttribute("books", books);
        model.addAttribute("pageNo", Integer.valueOf(pageNo));
        model.addAttribute("pageSize", Integer.valueOf(pageSize));
        return "admin.jsp";
    }
}
```

5. 视图层 admin.jsp

```jsp
<%@page language="java" pageEncoding="UTF-8"%>
<%@taglib uri="http://java.sun.com/jsp/jstl/core" prefix="c"%>
<%@taglib prefix="fn" uri="http://java.sun.com/jsp/jstl/functions"%>
<!DOCTYPE html>
<html>
    <head>
        <meta charset="utf-8"/>
        <meta name="viewport" content="width=device-width,initial-scale=1"/>
        <script src="js/jquery-3.6.0.min.js"></script>
        <script src="js/bootstrap.min.js"></script>
        <link rel="stylesheet" href="css/bootstrap.min.css"/>
        <title></title>
    </head>
    <body>
        <div class="container">
            <!--logo,网站名-->
            <div class="jumbotron">
                <h1>图书管理系统</h1>
            </div>
            <!--一个表单,提供用户选择要查询的内容-->
```

```html
<div class="row">
    <div class="col-lg-8">
        <form class="form-inline" role="form" method="get" action="admin-QueryAllBooks.do?title=title&type_id=type_id">
            <div class="form-group">
                <label for="title">书名</label>
                <input type="text" class="form-control" name="title" id="title" placeholder="书名">
            </div>
            <div class="form-group">
                <label for="type">类型</label>
                <select class="form-control" name="type_id" value="0">
                    <option value="0">请选择类型</option>
                    <c:forEach items="${listType}" var="type">
                        <option value="${type.id}">${type.typename}</option>
                    </c:forEach>
                </select>
            </div>
            <button type="submit" class="btn btn-default">查询</button>
        </form>
    </div>
    <div class="col-lg-4">
        <div class="col-lg-6">
            <p class="navbar-text navbar-right">${loginUsername}</p>
        </div>
        <div class="col-lg-6">
            <p class="navbar-text navbar-left"><a href="logout.do">注销</a></p>
        </div>
    </div>
</div>
<div class="row">
    <div class="col-lg-4"><a href="${pageContext.servletContext.contextPath}/insertBook.do" class="btn btn-primary" role="button">新增</a></div>
</div>
<!--展示全部图书的,以表格的形式展-->
<div class="row">
<table class="table table-dark table-hover table-bordered">
```

```html
        <thead>
            <tr>
                <th width="120">序号</th>
                <th width="100">封面</th>
                <th width="150">书名</th>
                <th width="100">作者</th>
                <th width="50">价格</th>
                <th width="100">出版社</th>
                <th width="80">出版时间</th>
                <th width="200">简介</th>
                <th width="80">操作</th>
            </tr>
        </thead>
        <tbody>
        <c:forEach items="${listBooks}" var="book">
            <tr>
                <th>${book.isbn}</th>
                <th><a href="queryBookByIsbn.do?isbn=${book.isbn}"><img src="${pageContext.servletContext.contextPath}${book.photo}" width="100" height="100"/></a></th>
                <th>${book.title}</th>
                <th>${book.author}</th>
                <th>${book.price}</th>
                <th>${book.publisher}</th>
                <th>${book.publishcationDate}</th>
                <th>${book.introduction}</th>
                <th><a href="queryBookByIsbn.do?isbn=${book.isbn}">编辑</a> | <a href="deleteBooksByIsbns.do?isbn=${book.isbn}">删除</a></th>
            </tr>
        </tbody>
        </c:forEach>
</table>
</div>
</body>
</html>
```

6. 运行项目

通过登录页面输入管理员的用户名和密码进入管理主页面,如图11.11所示。

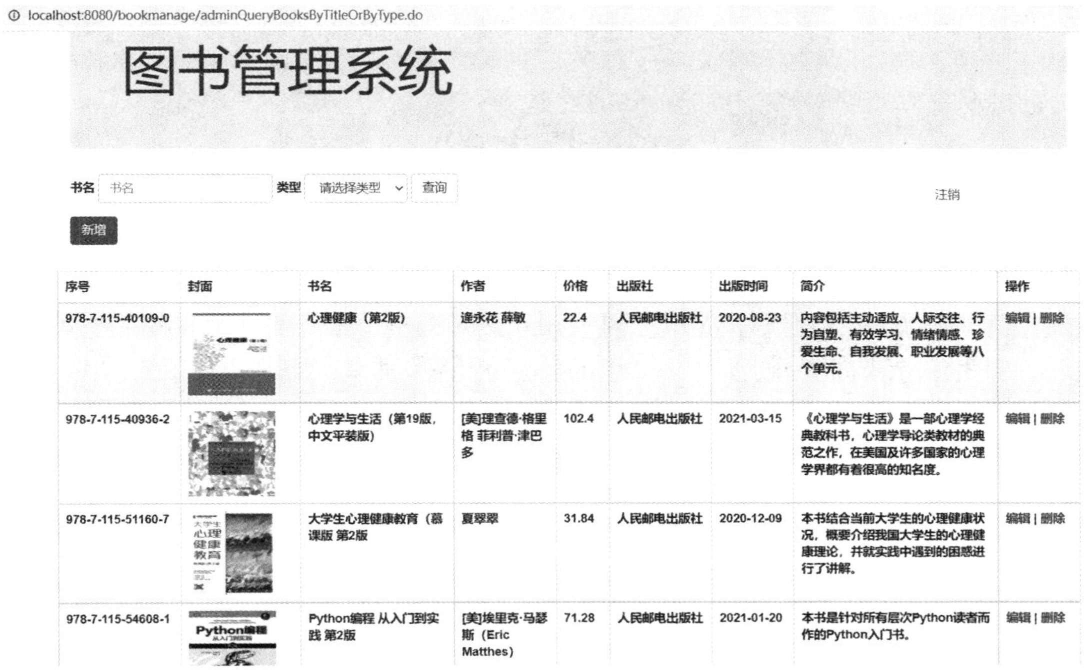
localhost:8080/bookmanage/adminQueryBooksByTitleOrByType.do
图书管理系统
书名
书名
类型
请选择类型
查询
注销
新增
序号 封面 书名 作者 价格 出版社 出版时间 简介 操作
978-7-115-40109-0 心理健康（第2版） 逄永花 薛敏 22.4 人民邮电出版社 2020-08-23 内容包括主动适应、人际交往、行为自塑、有效学习、情绪情感、珍爱生命、自我发展、职业发展等八个单元。 编辑 | 删除
978-7-115-40936-2 心理学与生活（第19版，中文平装版） [美]理查德·格里格 菲利普·津巴多 102.4 人民邮电出版社 2021-03-15 《心理学与生活》是一部心理学经典教科书，心理学导论类教材的典范之作，在美国及许多国家的心理学界都有着很高的知名度。 编辑 | 删除
978-7-115-51160-7 大学生心理健康教育（慕课版 第2版 夏翠翠 31.84 人民邮电出版社 2020-12-09 本书结合当前大学生的心理健康状况，概要介绍我国大学生的心理健康理论，并就实践中遇到的困惑进行了讲解。 编辑 | 删除
978-7-115-54608-1 Python编程 从入门到实践 第2版 [美]埃里克·马瑟斯（Eric Matthes） 71.28 人民邮电出版社 2021-01-20 本书是针对所有层次Python读者而作的Python入门书。 编辑 | 删除

图 11.11　管理主页面

参考文献

[1] ERIC J,RICARDO C N,IAN E,等.Java EE 6 开发手册·高级篇(第 4 版)[M].张若飞,丁永玲,李青,译.北京:电子工业出版社,2014.

[2] 贺智明,曾婕,王鹏飞.Java EE 企业应用开发技术[M].北京:清华大学出版社,2012.

[3] 刘彦君,金飞虎.Java EE 开发技术与案例教程[M].北京:人民邮电出版社,2014.

[4] 黑马程序员.Java EE 企业级应用开发教程[M].北京:人民邮电出版社,2017.

[5] 肖睿,肖静,董宁.SSM 轻量级框架应用实战[M].北京:人民邮电出版社,2018.